THE OBESOGEN EFFECT

THE OBESOGEN EFFECT

Why We Eat Less and Exercise More but Still Struggle to Lose Weight

Featuring a 3-Step Plan to Prevent and Counter the Effects of Silent Exposures

Bruce Blumberg, PhD

with Kristin Loberg

GRAND CENTRAL
Life & Style
NEW YORK • BOSTON

The information herein is not intended to be a substitute for medical advice. This publication is intended to provide helpful and informative material on the subjects addressed. It is sold with the understanding that the author and publisher are not engaged in rendering medical, health, or any other kind of professional services in the book. You are advised to consult with your health care professional with regard to matters relating to your health and in particular regarding matters that may require diagnosis or medical attention. The author and publisher specifically disclaim all responsibility for any liability, loss or risk, personal or otherwise, which is incurred as a consequence, directly or indirectly, of the use and application of any of the contents of this book.

Grand Central Life & Style
Hachette Book Group
1290 Avenue of the Americas, New York, NY 10104
grandcentrallifeandstyle.com
twitter.com/grandcentralpub

First Edition: March 2018

Grand Central Life & Style is an imprint of Grand Central Publishing. The Grand Central Life & Style name and logo are trademarks of Hachette Book Group, Inc.

Additional credits information is on page 287.

Library of Congress Cataloging-in-Publication Data
Names: Blumberg, Bruce, author.
Title: The obesogen effect : why we eat less and exercise more but still struggle to lose weight / Bruce Blumberg, PhD with Kristin Loberg.
Description: First edition. | New York : Grand Central Life & Style, 2018. | Includes bibliographical references.
Identifiers: LCCN 2017041727| ISBN 9781478970644 (hardcover) | ISBN 9781478970675 (ebook) | ISBN 9781478970668 (audio download) | ISBN 9781549168987 (audio book)
Subjects: LCSH: Obesity—Epidemiology—Popular works. | Endocrine disrupting chemicals. | Weight loss. | BISAC: SCIENCE / Life Sciences / Cytology. | HEALTH & FITNESS / Weight Loss. | MEDICAL / Toxicology.
Classification: LCC RA645.O23 B58 2018 | DDC 362.196/398—dc23
LC record available at https://lccn.loc.gov/2017041727

ISBNs: 978-1-4789-7064-4 (hardcover), 978-1-4789-7067-5 (ebook)

Printed in the United States of America

LSC-C

10 9 8 7 6 5 4 3 2 1

For Dejoie and Arielle

What if everything you thought you knew about the biology of body weight is ***wrong***?

Contents

PART II
In the Real World: How to Take Control in 3 Simple Steps

THE OBESOGEN EFFECT

INTRODUCTION

Fat: An Unrequited Love Story

When you think about the causes of overweight and obesity, conditions that now affect the majority of Americans, two factors likely come to mind immediately: dreadful dietary habits and lack of exercise. This is what I call the "orthodox wisdom" that we hear all the time. But what if I said you are wrong? Well, at least not 100 percent right. You're missing a huge influence that has been driving our epidemic for the last half century, and it has nothing to do with a penchant for sitting on the couch eating potato chips and drinking regular soda. It has to do with obesogens—chemicals in our environment that promote weight gain.

No one wants to be fat, but most of us are, despite working hard to eliminate unwanted pounds. Something is wrong with this narrative. I coined the term "obesogens" in 2006 to describe chemicals that can make you fat.[1] This sounded the alarm and spurred a flurry of scientific research studying the phenomenon of chemical-induced obesity. My team found that a chemical we were studying for other reasons had the ability to make mice fat. That started me thinking that there might be an alternative explanation for our irrepressible fatness other than calories in versus calories out. And I was right.

Take a moment to consider this from a purely logical standpoint: If weight were simply determined by calories eaten minus calories burned (more formally called the energy balance equation), don't you think we would be able to easily manage our weight? Why can we balance our bank checkbooks, but not our caloric checkbooks? In arithmetic, one plus one equals two no matter what language you speak. But one plus one can equal more than two when it comes to the weight equation of the human body. I will explain how this is possible in the book.

Observational studies in humans have pointed to a strong link between exposure to certain environmental chemicals and greater body mass index (BMI).[2] The BMI is a general measure that relates your weight in kilograms to your height.[3] BMI is often used as an indicator of obesity on one end of the spectrum and underweight on the other. In 1997, the World Health Organization (WHO) convened for its first meeting on the rising obesity epidemic and adopted new criteria for "normal weight" (BMI of 18.5–24.9), "overweight" (BMI of 25–29.9), and "obese" (BMI of 30 or higher).[4] The easiest way to measure your BMI is to use an online calculator, which will divide your weight in kilograms by the square of your height in meters to arrive at the number.

An important 2016 study showed that average BMI today is higher than it was a little more than a generation ago, even when our caloric intake and physical activity is about the same.[5] Put another way, adults today find it harder to maintain the same weight than did adults twenty to thirty years ago, even at the same levels of food intake and exercise. These days people are about 10 percent heavier than people in the 1980s, even if they eat and exercise as they did back in the heyday of leg warmers and Sony Walkmans. And despite what you hear, we exercise ***more*** than we did in the 1980s—not less. There is another explanation that will unfold chapter by chapter.

The time has come to present the untold story of obesogens with the hope that you can take better control of your waistline, your health, and especially the well-being of your children and future generations. After all, nowhere is the obesity epidemic more painfully disturbing to witness and acknowledge than in our young. In January 2016, the World Health Organization released a statement declaring that the number of obese children worldwide today is "alarming."[6] I will add the words "disheartening" and "unacceptable."

I don't mean to minimize poor diet and physical inactivity; these remain leading causes of overweight and obesity. But we in the scientific community are increasingly finding that exposure to chemicals in our diet and environment may be an underrecognized risk factor. In the last decade, other researchers and I have identified dozens of chemicals that can increase susceptibility to becoming obese in animals and trigger cells grown in a lab (cultured cells) to become fat cells.[7,8] The narrative about our obesity epidemic, which is now a severe public health crisis, continues to be stuck in the conversation about our modern lifestyles—too much food (especially the wrong kind) and not enough sweating. When doctors address patients who are overweight, they resort to the same old questions: "What are you eating? How much are you exercising?" Doctors rarely ask about what their patients may be exposed to (even unwittingly) in daily life. Indeed, many are hostile to the idea that chemical exposure may have effects on health. The quiz coming up shortly will begin to clue you in to what kinds of exposures

I am talking about, many of which may surprise you because you never knew there was a connection to your weight. Losing weight is no longer only about putting down the doughnuts and hopping on a treadmill.

What is missing from the never-ending conversation about weight and how to control it is the role played by obesogens and how these little-acknowledged yet potent and deleterious substances that we encounter daily—in our food, households, workplaces, and even medicine cabinets—are severely impacting our waistlines and overall health. Obesogens contribute to obesity by disrupting the normal development and balance of fat metabolism—how your body creates and stores fat. Obesogens can reprogram stem cells in the body to develop into more fat cells. Obesogen exposure also changes how your body responds to dietary choices and handles calories. So even though you have bought into the latest trends—Paleo, low-carb, gluten-free, Zumba, or CrossFit gyms—you can still struggle mightily with weight because of what is in your environment (broadly defined).

One of the most pernicious ramifications of obesogens is that their effects can be passed on to future generations. That's right: the effects of obesogen exposure can be ***heritable***. The havoc that obesogens wreak on our bodies can be passed down to our biological children, grandchildren, and beyond. This is why understanding the science of obesogens and knowing how to avoid them is particularly important for women who intend to become pregnant, are already pregnant, or have young children. The developmental years are a sensitive period in one's life, during which the body can be more vulnerable to, and affected by, chemical exposure. Our children, grandchildren, and beyond deserve to have the best possible chance to live long, healthy, and lean lives without being saddled with a predisposition to the burden of obesity and its related consequences. My hope with this book is to show you how to make it more likely that your children will succeed rather than fail to control their weight, as well as to help you reduce the impact of obesogens in your own life.

Now, before we begin the journey, I will share a little bit about me. I didn't start my doctoral life searching for obesogens. As happens with so many discoveries in science, I stumbled upon them while exploring other areas in biology. When I arrived at UCLA as a new PhD student in 1982, I wanted to study developmental biology—how organisms grow and develop from a single cell into complex, multicellular organisms such as humans. At that time, all of the developmental biologists at UCLA were exploring the genetics and development of the fruit fly, *Drosophila*, which I didn't find particularly appealing. Instead, I ended up studying the biochemistry of the extracellular matrix—the connective tissue that helps hold the body together. When I became a postdoctoral fellow at UCLA, I finally studied vertebrate developmental biology, looking for embryonic inducers—molecules that play critical roles in directing embryonic cells to form new tissues and tell the developing embryo where the head, arms, legs, and so on should go. Most everyone else in the field was studying peptide growth factors (proteins that stimulate cell growth) as embryonic patterning molecules, but my background with the extracellular matrix led me to look for small, fat-soluble molecules that could move freely through the sticky matrix around cells, whereas growth factors cannot. Small, fat-soluble molecules such as steroid hormones are ideal candidates for such molecules and were already known to be important for development. As you will soon learn, hormone levels can control how you behave and how your body develops and functions, even including your metabolism and risk for obesity. The receptors that these hormones bind to are powerful molecular switches that control the activity of many other genes. Since quite a few apparent receptors were then known for which no hormones had been identified (these are called "orphan receptors"), I set out to find new receptor-hormone combinations.

Later, I moved to the Salk Institute for Biological Studies in the Gene Expression Laboratory of Dr. Ron Evans, the world leader in hormone receptor research, to improve my chances of identifying new hormones. In 1996, while I was busily developing methods to identify new hormone-receptor combinations, I got a call from my colleague Dave Gardiner at the University of California–Irvine about deformed frogs that were being found in Minnesota. Dave and a few other scientists thought that the cause could be a retinoid (a chemical related to vitamin A) in the water and wanted to know how difficult it would be to test their hypothesis. I applied the same methods we developed to identify new hormones to find chemicals in the water that might be activating retinoid receptors to cause the types of deformities observed in the Minnesota frogs. That got me started on the path toward looking for environmental chemicals that disrupt hormonal signaling and thereby alter development (so-called endocrine disrupting chemicals, or EDCs). The rest is history that I will share in the book.

Fasten your seat belt. What you are about to learn will stun you on one level but inspire you on another (and make you feel a little better about yourself if you have struggled with your weight). You are not alone when it comes to matters of weight and the emotional toll of being on a seemingly endless diet to achieve and maintain weight loss. It ***is*** an uphill battle. You are also not alone if you answer "yes" to any of the questions coming up in the first chapter that will help you to grasp how many things in your environment could be making you fat—and keeping you fat despite your best efforts. Together we can identify the hidden factors that are sabotaging your health and weight loss efforts. Let's begin.

PART I

In the Fatlands: The New Science of Obesity

CHAPTER 1

Why Are We So Fat Despite Our Best Efforts?

The Hidden Factor No One Is Talking About

> If we could give every individual the right amount of nourishment and exercise, not too little and not too much, we would have found the safest way to health.
>
> —Hippocrates, the father of medicine (460–370 BCE)

If you could go back in time to an era when people believed the earth was flat, how difficult do you think it would be to convince someone that it was, in fact, round? What argument would you use to prove your point? And no, you cannot time travel with a globe in your pocket or images taken from the moon. Without any evidence of a round earth on hand, chances are you would have a tough time convincing anyone that you were right. After all, it is difficult to see the curvature of the earth from its surface—we see straight horizons, not curves. The idea of a spherical shape first appeared in Greek philosophy, credited to Pythagoras in the sixth century BCE. Aristotle provided evidence for it on empirical grounds by around 330 BCE, but it would take centuries for universal acceptance of a round earth that revolved around the sun. Even today, at least four U.S. professional basketball players claim to believe the earth is flat, so you cannot convince everyone, no matter how strong the data are!

When I speak before large professional audiences about obesogens and their impact on the obesity epidemic, I used to feel like that hapless individual trying to persuade the Flat Earth Society that the planet is indeed spherical. Old theories die hard, especially ones that people have clung to for a long time or when the reality seems to defy logic. This is absolutely true when it comes to our ideas about how and why we are so fat today. Laymen and doctors alike continue to believe that body weight is exclusively determined by the number of calories consumed compared with those burned—simple as that. Just look at the website of the National Institute of Diabetes and Digestive and Kidney Diseases (NIDDK), where it states definitively: "Overweight and obesity result from an energy imbalance."[9] If only this were the case. You might be surprised to learn that the number of health clubs has doubled over the same period that obesity has doubled. Clearly people are trying hard not to be obese. If only the solution revolved around diet and exercise alone.

Though most everyone has a clear picture of what obesity looks like, obesity is primarily defined by the body mass index (BMI). According to the WHO and American Medical Association, you are considered to be "underweight" if your BMI is less than 18.5, "normal weight" if your BMI is between 18.5 and 24.9, "overweight" if your BMI is between 25 and 29.9, and "obese" if your BMI is 30 or higher. A BMI beyond 40 is

considered to be morbidly obese and indicative of a serious health problem. The BMI is a calculation based on height and weight but does not distinguish muscle from fat and bone in the body. I often show a slide comparing three people with identical BMIs of 32. One has extensive subcutaneous (below-the-skin) fat, which is not really problematic for health; another has abundant visceral fat deep inside and around the organs, which, as we will describe later, confers many of the health risks associated with obesity. The third picture is of the "Governator," Arnold Schwarzenegger, at the pinnacle of his bodybuilding career. Despite having a

BMI of 32, he was not remotely obese or even overweight. Thus, BMI is not a perfect indicator of whether or not you are obese. However, it is easy to calculate and a convenient generalization.

Let's do one more imaginary time-travel experiment: Think about what it would be like for a family of Stone Age hunter-gatherers to experience the modern world (think *The Flintstones* meets *My 600-lb Life*). I picture stunned looks on the faces of our ancestors as they take in our modern conveniences—cars, cell phones, computers, electricity, and even supermarkets. And they are also taken aback by our body size. Given our twenty-first-century bulk, would they recognize us as members of their own species? Unlike us, they worked hard for every morsel and probably rarely had much leisure time to sit around eating unhealthy food. There were no fast-food restaurants or processed meals.

Hippocrates was right to stress the importance of moderation in diet and exercise as well as the value of food; diet was considered to be one of the most important interventions in Hippocratic medicine. This was certainly true in his time, but it misses a key part of the equation today. Hippocrates could not have known about endocrine disruptors or the effects they can have on our health and weight. And he, too, would probably be scratching his head in bewilderment at the obese masses at the dawn of the twenty-first century—a mere twenty-four hundred years after he lived.

Inexorably, we have grown accustomed to seeing a world of fat around us. The obesity facts are played over and over again in the media, so much so that they no longer surprise us. Overweight and obese individuals are the norm in most areas of the country instead of an anomaly. But fifty years ago you would have been hard-pressed to find many such people on the planet, let alone in one geographic location. As recently as 1990, obese adults in the United States comprised less than 15 percent of the population in most states. By 2010, thirty-six states had obesity rates of 25 percent or higher, and twelve of those had rates of 30 percent or higher.[10] Obesity has increased threefold over the past forty years and has doubled worldwide in the last twenty years; in the last thirty years, obesity has more than doubled in children and ***quadrupled*** in teens. Today there are more overweight and obese adults aged twenty years and older in the United States (70.7 percent) than those of normal, healthy weight and underweight.[11-13] More than one-third of U.S. adults are obese, and a whopping 86 percent of us are expected to be overweight by 2020. A full 20.6 percent of adolescents aged twelve to nineteen years are obese; 17.4 percent of children aged six to eleven years are obese; and almost 10 percent of kids only two to five years old are already saddled with obesity.[11,13] Individuals who had previously enjoyed life in the slim lane at the lower end of the body mass index are now gaining weight, too, and entering the "overweight" and "obese" categories. It becomes difficult to even see the problem because when everyone around you is overweight, it turns into the expected.

But this is not normal. Could it be that we really lack the willpower to control our waistlines through classic diet and exercise because we live in an age of abundance? That it is all about lack of self-control? I don't think so. Let's not forget that there is also an obesity epidemic among infants of six months of age and younger, an age group where food choices and limited physical activity cannot explain this phenomenon; infants move as little or as much as they want and eat until they are full. The obesity epidemic is not limited to humans, either—it is also affecting animals living with us, including domestic dogs and cats, feral rats living in cities, and, crucially, primates and rodents living in research colonies where diets are carefully controlled.[14] Are these animals also suffering from a lack of willpower? Highly doubtful.

The seminal 2016 study on our increased average BMI that I mentioned in the introduction was among the first to really call out the new facts.[5] The researchers in this particular study mined dietary data from 36,377 U.S. adults from the National Health and Nutrition Examination Survey (NHANES) between 1971 and 2008 and showed that we are 2.3 points higher on the body mass index today than we were just a generation ago—even after correcting for how much we eat and exercise. The published conclusion spoke volumes: "Factors other than diet and physical activity may be contributing to the increase in BMI over time." Granted, as I mentioned, BMI itself is not the best measure of obesity because it does not differentiate between fat mass and muscle mass or even where the fat is located, an important factor we will explore later. Yet despite the shortcomings of BMI as a metric, it does offer a reasonable reference point to use in evaluating body composition in the population.

So what is the most compelling explanation for these new facts? Something has changed in our environment—something that occurs where humans live, work, and eat.

HOW EXPOSED ARE YOU?

Take a look around you and think about a typical day, including that of your young children if you have them. In what ways can your environment impact your biology—and especially that of your children who are still developing, which makes them uniquely vulnerable? Here is a brief quiz that can help clue you in to the kinds of exposures that can cause a human body to create more fat. Many of these exposures could have happened long ago.

- Do you regularly drink from plastic beverage containers, including soda bottles, juice, and so forth?
- Do you have trouble sleeping well on a regular basis? (And yes, sleep deprivation can be considered an "exposure"; see chapters 5 and 9.)
- Do you live in an urban environment with air pollution or in a loud setting such as near a busy highway or airport?
- Do you feel that you have a lot of stress and/or don't respond well to stress?
- Do you buy a lot of prepackaged foods/meals rather than whole foods and fresh ingredients from which to create meals?
- Do you eat conventionally grown fruits and vegetables?
- Do you drink tap water?
- Are you taking drugs that list weight gain as a potential side effect?
- Do you clean your home with commercial, non-green products?
- Are you exposed to herbicides, pesticides, and fungicides around your home or neighborhood?
- Do you use air fresheners or scented products (including cosmetics that list "fragrance" in their ingredients)?
- Do you spend most of your days being inside, in a car, in an office?
- Do you sit more than three hours per day?
- As a child, did you drink a lot of juice, soda (regular or diet), and flavored milk or eat a lot of processed foods?
- Did your mom smoke when she was pregnant, or were you exposed to secondhand smoke as a child?
- Was your mom overweight or obese while she was pregnant with you?
- Was your dad obese before you were conceived?

Don't panic if you answered "yes" to any of these questions. Knowledge is power, and I will equip you with the information you need to take better charge of your waistline and health. In this chapter, we will take a tour of the obesity epidemic so you have a solid foundation to understand what I am calling "the new science of fat." It will help you begin to see why some people gain weight faster and lose it more slowly than others, irrespective of diet and exercise.

THEN AND NOW: THE MYSTERY OF OUR MODERN SCOURGE

The image below shows two different individuals.

Back in our hunter-gatherer days more than ten thousand years ago, when we had to forage for our food by collecting wild plants, fishing, and pursuing wild animals, most of us resembled

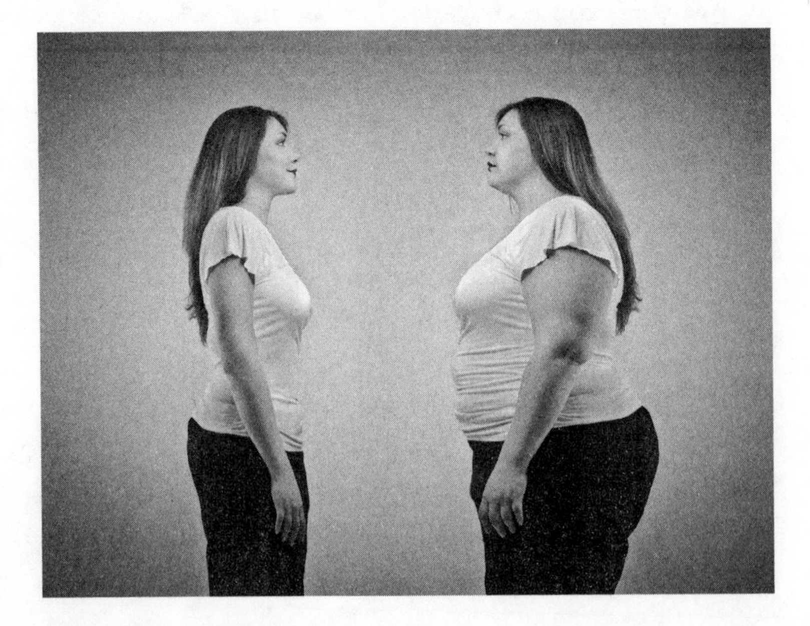

the individual on the left. Modern hunter-gatherers in Africa and elsewhere who continue to live this way still look like the woman on the left. However, many of us come closer to looking like the person on the right.

Some would argue that one of the reasons it is so easy to turn into the person on the right is that we evolved when food was much harder to find. We were programmed for survival in a world where food was scarce and periods of famine were frequent. But today we live in a calorie-rich environment 365 days a year. Food is abundant, and unhealthy foods tend to be the most affordable, available options. Many of these unhealthy foods are engineered to act on our brain's pleasure centers to keep us addicted and coming back for more, similar to how drug addicts on heroin or crack cannot easily quit.[15] While the theories around whether obesity-causing genes are widespread, or whether there is an obesity-favoring metabolic program, the idea that we have been molded by evolution to use calories sparingly makes some sense and is worth considering.

The "thrifty gene" hypothesis suggests that the two images on page 18 reflect an obvious mismatch between our Stone Age genes and space age circumstances. Having access to ample calories allowed the body to build fat stores quickly when food was available so it could endure long stretches of food shortages, which were inevitable and common throughout much of human evolution. Perhaps we retain essential components of this hunter-gatherer genome—after all, a small percentage of each of our genomes is derived from our Neanderthal and Denisovan cousins.[16] These genes helped us store fat during times of abundance so that this fat could be used during those regular periods of scarcity, and thus they lead to energy being spent sparingly. The end of famine in developed countries means that accumulated changes in genes that were once favorable for storing fat instead could cause widespread obesity. In other words, "famine genes" that were once advantageous for survival became liabilities that could threaten health and longevity in the land of plenty.

Geneticist James Neel first described the "thrifty gene" hypothesis in 1962 to help explain why diabetes has such a strong genetic basis yet results in such negative effects.[17] According to the theory, the genes that predispose someone to diabetes—the "thrifty genes"—were advantageous long ago. But once modern society changed and we developed agriculture, our access to food changed, too. Our bodies no longer needed those thrifty genes, but they were still active. Today they continue to prepare us for a famine that will not arrive. Our thrifty genes are partially responsible for the obesity epidemic, which is closely tied to the development of diabetes.

If there are such things as "thrifty genes," then it must be possible to identify them with the advent of modern, affordable DNA-sequencing methods. This is where the thrifty gene hypothesis has run into some critics. If these thrifty genes have been around for the two hundred thousand years or so that we *Homo sapiens* have existed, and agriculture began only around twelve thousand years ago, then pretty much everyone should carry the majority of the thrifty genes that have ever existed. Dr. John Speakman, a prominent challenger of the thrifty gene hypothesis, showed in a 2016 paper that none of the common obesity-related genes that had been identified conferred any properties or traits that could be considered to have provided an adaptive advantage.[18] On the other hand, perhaps there are bona fide thrifty genes that have yet to be identified, so the case is not yet closed. As we will discuss in chapter 5, it is clear that there is a "thrifty phenotype." That is, in both human and animal studies, it is possible to identify individuals that store more of the calories they consume as fat, whereas others burn more calories or eliminate more in their feces. If thrifty genes are not responsible, then what is?

I am sure that you know people who can eat large volumes of food at meals yet remain remarkably lean. But consider this: They would not have lasted long in the Stone Age. These people we view as fortunate today have the opposite of a thrifty phenotype. The person who eats a sandwich and gains weight does so by retaining more of those calories. He will survive much longer under conditions of starvation due to the excess energy his body stored in the form of fat. This is only viewed as a negative today because instead of facing starvation, we live amid abundance (at least of calories) and can easily end up looking like the individual on the right on page 18. In our culture, that is neither attractive nor healthy in the long term.

Every week I encounter skepticism about the biology of weight gain and loss. For decades, we have been told that our weight is a black-and-white reflection of the energy balance equation. If you take in more than you expend, you will gain weight; and if you burn more than you consume, you will lose weight. This rationale has always sounded so plausible and commonsensical. It is true to some extent when dieting. But the "calories in/calories out" model is obsolete. People want so badly to believe in this trivial explanation of metabolism, but if it were that simple and straightforward, then we would not be as overweight and obese as we are. As I posited earlier, why can we easily balance our financial but not our caloric checkbooks? Too much food and too little exercise does not fully account for the extent of our obesity. It fails to explain all the factors at work; it does not consider all of the contributing factors to obesity, including obesogens, which we will talk about in the following chapters.

Before I get to the conversation about obesity-causing chemicals, we should cover a few fundamental facts that need to be clarified.

A CALORIE IS NOT A CALORIE

Ask a group of people what a calorie is and most will tell you it is "the thing in food that makes me fat." Well, not really. A calorie is a unit of heat energy, specifically the amount of heat required to raise the temperature of 1 gram of water 1 degree Celsius (about 1.8 degrees Fahrenheit). What we in the United States commonly call a calorie is, in fact, equal to 1,000 calories, or 1 kcal (as you will see on food packaging in Europe). Calories in food are measured using what is called a bomb calorimeter. Bomb calorimeters determine heats of combustion by igniting a sample in pure oxygen at high pressure in a sealed vessel and measuring the resulting rise in temperature. However, people—including many scientists and medical doctors—misunderstand that the human body is not like a bomb calorimeter.

The human body does not burn calories the same way a bomb calorimeter does. A calorie is a calorie only if you are trying to heat your house. You have probably heard that protein and carbohydrates give approximately the same amount of energy, about 4 kcal per gram, while fat provides 9 kcal per gram. Fats are more energy dense. But biologically speaking, proteins, carbohydrates, and fats are vastly different because each gets metabolized differently and affects us in separate ways. The way your body responds to your eating 100 calories of carbohydrates in the form of refined sugar (about 4 heaped teaspoons of the white stuff) will not be the same as if you consumed 100 calories of pure fat in the form of butter (about 1 tablespoon). You can even feel the difference in hunger cues and levels of satiety. Want an example? You can do an experiment yourself: one day, have a bowl of cereal with fresh fruit and nonfat milk and see how long you can go without feeling hungry and eating again. The next day, try an egg-plus-cheese breakfast that contains the same number of total calories as the cereal breakfast. The carb-heavy cereal breakfast will leave most people hungry within about ninety minutes. The fat- and protein-heavy egg-and-cheese plate will likely keep you satisfied for hours. Clearly, the body didn't use—or experience—each breakfast in the same way, even though each contained the same amount of "energy." Why?

A lot is going on when you eat. You are not just consuming food. Eating triggers a multitude of hormonal pathways in the body that impact how food gets processed, how your brain interprets satiety signals, and how you ***feel***. Later in the book, we will go into all these details about how food affects the body and whether or not it leads to weight gain or loss. But for now, I want you to begin to change how you perceive calories.

If everyone used calories the same way, then you would never see a difference between person A and person B in their ability to maintain a certain weight with the same dietary intake and physical activity. Just balance the caloric checkbook and that's that. But just as distinct foods trigger different processes within the body, each individual body does not process identical foods in the same way. Nutrition labels lead to a "one size fits all" mentality. The "calorie is a calorie" dogma assumes that everyone uses calories equally, which is far from the truth. Another way to quickly grasp the gap between how calories behave in a vacuum and how they affect a human body is to consider a 2015 study that challenged a lot of what we thought we knew about "healthy" food.

Eran Segal, Eran Elinav, and their colleagues at the Weizmann Institute of Science in Israel homed in on one key component often considered in designing balanced diet plans: the glycemic index, or GI for short.[19] The GI was developed decades ago as a metric for how foods, particularly carbohydrate-containing foods, impact the amount of glucose in the blood. The GI uses a scale of 0 to 100, comparing foods with the reference point of pure glucose, which has a GI of 100. Foods with a high GI, such as potatoes, crackers, and white bread, are quickly digested and absorbed, causing rapid but transient elevations in blood sugar, which in turn trigger a spike in insulin, the hormone responsible for ushering glucose out of the bloodstream and into cells for use or storage. Insulin also stimulates fat and amino acid uptake into cells and inhibits the body from breaking down stored fat, glycogen, and proteins. Lower-GI foods such as whole grains, rice, black beans, and some starchy vegetables are digested more slowly, producing a gradual rise in blood sugar and insulin levels. Low-GI foods hardly change blood sugar levels at all.

GI values have long been assumed to be fixed numbers, intrinsic to the food itself. If this were the case, then everyone would respond to the same food in the same way. Segal and Elinav's team conducted an elegant experiment to test whether this was the case or not. First, they recruited eight hundred healthy and prediabetic volunteers aged eighteen to seventy years and gathered data through health questionnaires, blood tests, body measurements, glucose monitoring, and stool samples. Using a mobile app, the participants recorded food intake and lifestyle information. Everyone was instructed to eat a standardized breakfast that included foods such as bread each morning.

Their study analyzed the responses of participants to a total of 46,898 meals.[19] What they found shook the foundations of the diet field. As expected, age and body mass index appeared to impact blood sugar levels after meals. Surprisingly, people showed radically different responses to the same food! The GI of any given food was not a set value; rather, GI could differ depending on the individual. Person A, for instance, ate tomatoes and experienced an immediate spike in blood sugar levels, whereas person B ate the same tomatoes but did not experience an equivalent surge in blood sugar. The researchers concluded that customizing meal plans to personal biology may be the future of dieting,[19] and I could not agree more. While the concept of personalized medicine is widely accepted with respect to how pharmaceutical drugs affect individuals differently, the same doctors who readily accept that people respond differently to the same drug are much less willing to believe that people can respond differently to the same diet. One size does not fit all. We will also see in chapter 5 that a widely underappreciated factor in how people respond to foods is their gut microbiome—the microbial communities that live in the stomach and intestines and play important roles in how we metabolize food and even whether we feel hungry or not.

No two metabolisms are identical. This is strikingly evident when you consider that about 20 percent of type 2 diabetics are of normal weight and that some people who are obese never develop metabolic diseases and obesity-related conditions such as high blood pressure and unfavorable cholesterol profiles.[20] Therefore, while obesity and type 2 diabetes are not inextricably linked, most type 2 diabetics are obese, and most obese people are at risk for type 2 diabetes, among other conditions. But it is not so simple as obesity causes type 2 diabetes. Let's consider some of the other myths swirling around obesity.

IT MUST BE GENETIC, RIGHT?

The next time you are in a public venue surrounded by lots of people, take a look around you and note the different body types. You will likely find that there are roughly three different, dominating patterns in how people are built physically. In the 1940s, the American psychologist William Sheldon and colleagues proposed that people could be grouped into three different body types ("somatotypes"), named after the three germ layers—the first major types of cells that are formed in the early embryo.[21] The "ectomorph" tends to be thin and slight in stature; he or she does not appear to have a lot of muscle or fat (but according to Sheldon has a more developed nervous system). The "endomorph," on the other hand, is the person who is described as rotund with more fat mass than well-defined muscle mass. The "mesomorph" is somewhere in the middle with an overall lean body and defined muscles.[22] Sheldon went too far by assigning psychological characteristics to these physiques—the round endomorph being fun loving and the thin ectomorph being anxious and intense.[22] This idea was about as reliable as attributing personality traits to the signs of the zodiac. Nevertheless, the popularity of the body type classification has remained, because like some stereotypes, it contains a kernel of truth: the somatotype designations recognize that a person's propensity toward building muscle or storing fat is largely predetermined.[23]

This leads to an important question: How much of obesity is

genetic? While the hereditary origins of obesity have long been assumed, a genetic contribution to obesity became evident only in the last two decades. Evidence from twins and animal studies once suggested that genetic factors account for 40 to 70 percent of the variation in BMI.[24] But recent studies put this number much lower, at perhaps only 20 percent.[25] Although several single genes are linked to obesity, there is no "obesity gene" or even a group of obesity genes that accounts for a major part of the obesity epidemic. Most geneticists believe that obesity is polygenic. That is, it would take changes in many genes (largely those associated with appetite and metabolism) to produce the obesity epidemic we are witnessing. But these kinds of changes could not possibly have happened worldwide over the past forty years. On the other hand, many people who carry these genetic variants linked to increased BMI are not obese; therefore, it is highly likely that other factors interact with any such genetic predispositions.[26,27] One such factor is the environment, and we will see in a later chapter just how powerful environmental forces can be to change the expression of a variety of genes in your body—without making any changes in your personal genetic sequence. Not only can you be born with a tendency to gain weight easily due to environmental influences in utero, but the expression of your genes throughout your entire life can be changed by environmental factors that can make either weight gain or weight loss easy.

YOU MUST NOT BE EXERCISING ENOUGH

The prevailing wisdom, I call it "the couch potato syndrome," is narrow-minded and largely wrong. Although it is commonly stated that we are more sedentary today than in years past, that is debatable—the facts don't fully support this point of view. Just as there are studies showing less physical activity[28] today than in generations past, there are other studies revealing that we are in fact ***more*** active today.[5] By some reports, overall inactivity has dropped in recent years and participation in sports, fitness, and related physical activities has increased. When I think about my own life, I find that adults try to be more active today than the adults I knew growing up. I can't recall that there were very many health clubs when I was young. Today, many people strive to stay fit and keep moving, which is reflected by the number of gyms, the popularity of online fitness programs, and the increasing number of fitness apps and fitness tracking devices (for instance, Fitbit and Jawbone). People are trying hard to get into shape.

No matter what your position on the value of exercise (and I think it is beneficial in very many ways), even inactivity coupled with poor diet cannot explain the magnitude of the obesity epidemic. In addition, exercise is linked to weight gain rather than weight loss in many studies. A provocative 2014 study showed that a substantial number of women who take up an exercise regimen wind up heavier afterward than they were at the start, with the weight gain due mostly to extra fat, not muscle.[29] Although exactly why this happens is not completely clear, one of the reasons is likely to be physiological: when you exercise more, your appetite naturally increases.[30] Feeling hungrier can cause you to crave carbohydrates to restore the glycogen depleted from your muscles, which could cause you to gravitate to unhealthy snack foods containing lots of refined sugars, fats, and salt. Counterintuitively, you might also move less at times when you are not exercising (perhaps because you are sore). For some people, especially those who are already obese, rigorous, challenging exercise can cause measurable and significant neural activity in brain regions responsible for food reward and craving.[31,32] Such brain activity can make it nearly impossible to say no to, and not overindulge in, tasty snacks high in fat and sugar. Interestingly, studies investigating the brain activity of lean, fit people show their food-reward centers respond less aggressively to images of delectable foods[33] (but more aggressively to low-calorie foods).[34,35]

Another explanation is psychological: when you know that you are burning more calories by exercising, you tend to give yourself greater permission to eat more. Ultimately, this could make you overcompensate for the number of calories lost through physical activity.

To be sure, exercise serves an important role in your overall health. Although you can't rely on exercise alone to promote weight loss or as the ultimate antidote to weight gain, physical activity is useful in reducing the risk of developing heart disease, diabetes, dementia, and many other conditions.

YOU EAT TOO MUCH, DUMMY

Diet trends come and go. Sometimes, low-fat is all the rage; other times, it's low-carb, Paleo, or Atkins-type diets. There is no doubt many of us consume too many carbohydrates, especially in the refined, sugary form. This was a natural consequence of the "war against dietary fat" that began with the McGovern Report, which first established Dietary Goals for the United States, published in 1977.[36] Unfortunately, since fat often helps to make food tasty, a substitute ingredient was needed, so—you guessed it—sugar was added to food to replace the tasty fats. Take a walk down the candy aisle in any supermarket and count how many types of candy say they are low-fat or zero-fat foods, implying that they are therefore healthy. We also gravitate toward salty, fatty foods with added sugars that provide little nutritional value, titillate our taste buds, and apparently increase food consumption.[37] But even these dietary shortcomings do not explain our weight problems. If they did, then focusing on nutrient-dense foods and steering clear of classic fattening foods such as fried chicken and doughnuts would cinch our waistlines. Adhering to the strictest, "best" diet does not necessarily guarantee that the weight will melt away and you will stay lean for life.

Is there another explanation for people who eat well and cannot lose weight? Just look at the example of obese people who have fought the good fight and lost hundreds of pounds. Why do more than 83 percent of these people who have worked and sweated to lose large amounts of weight gain it back within a few years?[38,39] How does the weight loss industry—comprising diet companies, nutritionists, supplement manufacturers, pharmaceutical giants, diet book authors, fitness trainers, lifestyle experts, "detox" juice bars, obesity doctors, and gastric surgeons—rake in multiple billions every year without offering a bulletproof cure?

My answer, in short, is: obesogens. Clearly, other factors play into the risk for obesity, such as chronic sleep disruption and stress, but we will get to these other fat-inducing factors later in the book. For now, we are going to focus chiefly on the growing body of evidence pointing to environmental substances called obesogens as additional, under-acknowledged factors that alter metabolism and predispose some individuals to weight gain.

THE OBESOGEN HYPOTHESIS

In 2002, Dr. Paula Baillie-Hamilton of the United Kingdom wrote the first article relating environmental chemicals to obesity. She had a pet theory: exposure to toxic chemicals was preventing her from losing weight after having four children. Her article, "Chemical Toxins: A Hypothesis to Explain the Global Obesity Epidemic," was a review of published toxicological studies dating as far back as the 1970s.[40] Baillie-Hamilton showed a correlation between the rise in chemical manufacturing after World War II and the rise in obesity, claimed that chemicals caused obesity, and declared that purging the chemicals would reverse the process. However, her most important contribution went unnoticed by most readers. This was that exposure to high doses of a variety of environmental chemicals, including some pesticides, solvents, plastics, flame retardants, and heavy metals, was shown in some studies to lead to weight loss, whereas animals exposed to low doses of the same chemicals ***gained*** weight. This seemingly small yet significant fact was ignored because the toxicologists who conducted these studies typically were concerned only with weight loss—a prime indicator of toxicity. Moreover, the idea that a chemical could have one effect (weight gain) at a low dose and the opposite effect (weight loss) at a high dose did not fit with the orthodox "dose makes the poison, toxicity is linear" worldview of many toxicologists. Baillie-Hamilton's work went largely unnoticed because it was published in an obscure place, the *Journal of Alternative and Complementary Medicine*, which is not read much by researchers.

Another reason Baillie-Hamilton's work did not receive much attention was that she had detected only a ***correlation*** between chemicals and obesity and was overselling the story. You can correlate many things with the increase in obesity. In my talks, I sometimes humorously point out a number of absurd correlations between the rise in obesity and other factors. These include such things as the number of gyms/health clubs (which, as I mentioned earlier, have doubled in step with the doubling of obesity in the United States), the number of SUVs on U.S. highways, or even the number of dermatologists. While it can be entertaining to link two seemingly unrelated phenomena just because you can, correlation is not causation. Gyms, driving an SUV, and dermatologists probably do not cause obesity. Many, if not most, correlations are entirely coincidental. As scientists, it is our job to identify causation, not just correlations.

One person who did notice Baillie-Hamilton's paper was Jerrold (Jerry) Heindel at the National Institute of Environmental Health Sciences (NIEHS). Jerry was a program officer—an agency official who is responsible for managing a portfolio of grants to ensure that the research being funded fits the mission of the agency, in this case to understand how the environment (broadly defined) influences health. Upon reading the Baillie-Hamilton article, Jerry made the intellectual leap that it was not all toxic chemicals per se that were responsible for obesity, but rather endocrine disrupting chemicals (EDCs), which are chemicals that can interfere with how hormones work in our bodies (much more on this later). He then set out to drum up interest in this new field among scientists already working on EDCs. Jerry wrote an influential article, "Endocrine Disruptors and the Obesity Epidemic," which appeared in a mainstream journal, *Toxicological Sciences*, in 2003, placing endocrine disruptors and obesity on the same map for the first time.[41] He is too modest to admit it, but Jerry is really the father of the environmental obesogen hypothesis. I often jokingly tell waitresses that he is my dad when we travel to meetings around the world. Sadly, Jerry retired from NIEHS at the end of 2016. The community of researchers working on EDCs will greatly miss his infectious optimism and influence and his propensity for calling BS on any conclusions not strongly supported by high-quality data. Jerry had a very strong, positive influence on the field that was unique among NIH program officers.

In 2006, my team and I published our now classic paper identifying tributyltin (TBT) as an environmental chemical that leads to fat cell development and weight gain.[42] TBT is a chemical (specifically, an organotin) that kills fungi and was once widely used in antifouling paints on ship hulls but continues to be found in seafood and can be found as a contaminant in vinyl plastics. We coined the term "obesogen" to describe these fattening chemicals. Obesogen exposure changes how energy is stored and used in the body to favor weight gain and obesity. We proposed that obesogens can act directly on the fat to produce more fat cells, larger fat cells, or more cells that are destined to become fat cells. We suggested that obesogens can also derail the normal processes that the body uses to maintain weight, regulate appetite, and control metabolism, causing exposed individuals to be increasingly susceptible to weight gain despite normal diet and exercise. Work in my lab and in other labs around the world has shown all of these mechanisms to be at play.[7] Limiting exposure to obesogens is not just about reducing risk for overweight and obesity. Reducing exposure to endocrine disrupting chemical obesogens will reduce your risk for various illnesses—from heart disease to cancer[43,44] and even dementia.[45] In upcoming chapters, you will gain an understanding of these mechanisms together with some insights into how you can avoid more damaging effects from fat-inducing chemicals.

WHERE DO OBESOGENS LURK?

To be sure, not all EDCs are obesogens, and not all obesogens are the kinds of nasty toxic chemicals we think about, such as those from a smoke-belching factory exhaust pipe. Obesogens can come from sources such as food (including fruits, vegetables, and your favorite homemade comfort foods), beverages (including water), pharmaceuticals, and everyday household items we use routinely, such as personal care products, cosmetics, sunscreens, shower curtains, air fresheners, mattresses, clothing, kitchen tools and plastic storage containers, canned goods, toys, furniture, flooring, and cleaning supplies. We have currently identified about fifty potential obesogens,[7] but there has been no systematic attempt to identify obesogens, and there are more than one thousand known EDCs (despite no systematic program to identify them either). Therefore, I expect that more research will discover hundreds of new obesogens. Following is a general list of known obesogens to date:

Chemical pesticides: Commonly used pesticides in conventional farming, especially the breakdown product of DDT (DDE), have been definitively linked to increased BMI in children.[46-48] In laboratory settings, they cause rodents to become resistant to insulin, which can lead to diabetes. While DDT is banned in most of the world (but is still used in many parts of Africa), other obesogenic pesticides exist and are present in conventionally grown produce and can contaminate tap water.

BPA: Found in plastic bottles, can linings, medical devices, and cash register receipts, bisphenol A (BPA) has been shown to cause fatty liver as well as increased abdominal fat and glucose intolerance in animals.[49] Early studies show that replacement chemicals such as BPF and BPS have the same activity as BPA.

PFOA: Perfluorooctanoic acid (PFOA) is used in nonstick coatings such as Teflon, food packaging (microwave popcorn bags), backpacks and luggage, carpeting, and clothing and has even been found in water.[50] Studies are still under way, but the current thinking is that such chemicals are obesogens, among their many other detrimental effects.

Phthalates: Phthalates are ubiquitous—they are found in plastic and vinyl, including medical devices and tubing, toys, wall paint, air fresheners, as well as in numerous beauty and personal care products. Some phthalates have been classified as carcinogens and have been shown to be obesogenic in animal studies and linked with obesity in humans.[7]

TBT: My laboratory favorite obesogen, tributyltin (TBT), is a powerful fungicide and bactericide that was once widely used in antifouling paints on ship hulls. TBT still contaminates seafood and can be found in vinyl plastics as an unintended contaminant of other organotins used as heat stabilizers and as a wood preservative.[42]

PCBs: Polychlorinated biphenyl chemicals were once used as electrical insulators and flame retardants until they were banned in 1979. But they still persist in the environment, and exposures typically come from contaminated fish, meat, and dairy products.[47,48] PCBs are obesogens and are also known to cause diabetes.

PBDEs: Polybrominated diphenyl ethers are flame retardants found in a lot of furniture, particularly foam. They are also found in cars, electronics, building materials, plastic foams, and textiles. Many have been banned, but others are still in use. PBDE replacements such as triphenylphosphate and tetrabromo bisphenol A are also obesogens.[7]

Soy: A common feed for livestock, soy contains high levels of phytoestrogens, many of which can be obesogenic when consumed by infants or children. Perinatal exposure to most types of estrogens, including soy, can promote the growth of fat cells and obesity later in life.[51]

MSG: Used to make foods taste better, monosodium glutamate (MSG) first made the news as the suspect in "Chinese restaurant syndrome," which in sensitive people can cause flushing, headaches, tachycardia (rapid heartbeat), and excessive sweating. Most people lack this sensitivity to MSG, but it is also an obesogen that is found in many canned and packaged foods, soups, and a variety of restaurant foods.[52]

HFCS: You likely already know that high-fructose corn syrup (HFCS), a refined sugar, is not the same as real sugar. We will be going into detail about why sugar is so obesogenic. The conversation about sugar can be complex, and much has been written about it (see the works of Robert Lustig and Gary Taubes, among others). Added sugar, natural or otherwise, is often considered to be "toxic" now, but what does that mean? And what is the difference between glucose, which serves as fuel in the body, and fructose, which is processed almost entirely by the liver, where it is stored as triglycerides? In 2014, a study by Michael Goran and his colleagues at the Childhood Obesity Research Center at the Keck School of Medicine of USC found that the levels of fructose in many popular beverages and juices were much higher than expected, so high that they could increase one's risk not just for obesity, but also for diabetes, cardiovascular disease, and liver disease.[53] They went on to show that fructose levels in breast milk were positively associated with body fat in infants.[54]

Artificial sweeteners: Swapping your regular soda for diet will not make you thin. The same goes for diet and low-fat foods that are filled with artificial sweeteners. Even ones marketed as "natural" can be problematic. When we cover the microbiome, you will see what these chemicals do to the gut bacteria, which in turn adversely affects your metabolism.

Nicotine: The main chemical found in cigarette smoke encourages obesity in the children of mothers who smoked when they were pregnant. In fact, one of the first connections made between obesity and human fetal development came from studies of exposure to nicotine in utero. Smoking has decreased over the past generation, but nicotine by-products can still be found in the blood of nonsmokers, attributable to secondhand smoke. More on this later.

Air pollutants: Many chemicals found in polluted air, especially in urban environments, are obesogens. See chapter 9 for more details.

Parabens: These are a class of obesogens known to disrupt hormones by mimicking estrogen. They are used as preservatives in many personal care and beauty products such as cosmetics, lotions, and hair products.

Two questions that scientists have been trying to answer are: (1) How much does obesogen exposure contribute to the obesity epidemic? And (2) What is its cost to society? My colleague Professor Leo Trasande from NYU is a specialist at such calculations. Leo is a pediatric epidemiologist who has led several teams of experts in brainstorming exercises intended to estimate the cost of EDC exposure in particular to the disease burden in the United States and in the European Union.[55,56] These efforts used very conservative estimates for how certain we were about exposure-disease linkages, the quality of the evidence, the types of studies they were based on, and so forth. The numbers are sobering: €23.9 billion in the EU and $5.9 billion in the U.S. for exposure to only three obesogens (bisphenol A, diethylhexyl phthalate, and the pesticide DDT) per year. These were the only obesogens (out of at least fifty known chemical obesogens) for which sufficient data were available to make the calculations. Therefore the actual economic impact of obesogens is probably much higher.

The hardest fact to accept is that it can be very difficult, if not impossible, to lose weight permanently, due at least in part to the effects these chemicals have on the body. Unlike a computer program that can be debugged, human programming does not work like that. Our bodies are self-regulating for the most part (temperature, blood pressure, hormonal cycles, and the like), and the biological control mechanisms for these can be thought of as a type of "programming." I am amused and also somewhat frustrated when I see diet books claiming that their protocol can help you to "reprogram" or "rewire" your fat cells and "make over" your metabolism. In reality, those are very difficult goals to achieve—especially in the continual presence of obesogens. These chemicals are reprogramming the body in ways that are not easily reversible. But with knowledge and strategy, it is possible to protect yourself and your family, particularly your young children, from exposure. We will start by establishing a solid understanding of how obesogens can program the body to store more of the calories you consume as fat and contribute to weight gain that is difficult to overcome.

CHAPTER 2

A Modern Chemistry of Fat

How We Gain and Lose Weight in the Twenty-First Century

Nearly everyone took biology in high school, many in college and beyond. But most people—even those with science degrees—would probably fail a test today on the biochemistry of fat in the human body given all the latest science that has emerged over the past few years. It is breathtaking on one level and heartbreaking on another. In this chapter, I will tell that tale, starting with how a healthy, normally functioning body controls weight, and then move into the new science of obesogens and fat.

THE COMPLEXITY OF METABOLISM

The diet community will have you believe that we know enough about the human body to design perfect weight loss programs that easily and permanently reduce anyone's weight. Some (for instance, Weight Watchers) are based on the "calories in versus calories out" model—just reduce total caloric intake and you will be slimmer. The Atkins-style diet and its Paleo relatives allow as many calories as one wants, so long as very few of them are from carbohydrates (and most of these from vegetables). The Pritikin Diet and its derivatives (such as Ornish and South Beach) allow abundant complex carbohydrates but little to no fat.

Surprise: No one really knows exactly how all the inner machinery of the human body functions, particularly when it comes to weight gain and loss. This is especially true on an individual basis, as no two bodies are identical. If we knew exactly how everything worked, don't you think that we could easily solve most problems in medicine today and prescribe a fail-safe weight loss program for you? Would there be any need for the colossal ($60 billion a year and growing) diet industry?

Assumptions about biology circulate like rumors. No one, for example, has challenged assumptions with respect to individual differences. How do some people gain weight under their skin (subcutaneous fat) while others pack on the pounds viscerally—around their intestines (which is very dangerous metabolically)? How these controls are set is not one-size-fits-all, which helps explain why there is no such thing as one diet (other than starvation) that works on everyone. Why do *The Biggest Loser* winners almost always gain back the weight they fought so hard to lose?[57]

In the previous chapter, I covered the main different body types you commonly see. There are people who seem to be naturally thin, those that are large, and individuals who are in between the two extremes ("average"). By the time you are an adult, your body has a natural tendency to maintain a certain weight and will adjust its internal processes accordingly. We call this the metabolic set point theory. This set point determines how much food you need to eat to gain weight, how much exercise you must do to lose weight, your propensity to food cravings and addiction, and how much you are willing to work for food. The set point programming for weight is not just genetically determined; it can be influenced by environmental forces and is largely set during critical periods of development, from in the womb through puberty. In the words of my friend and colleague Jerry Heindel, "A good start lasts a lifetime." And a bad start sets an individual up for a lifelong higher risk of experiencing health challenges of all kinds—obesity among them.

An example of another set point is body temperature: if temperatures go above or below 98.6 degrees Fahrenheit (for instance, from exposure to a hot environment or an infection), a variety of physical mechanisms are activated in an attempt to bring the body temperature back to normal. Does this happen in the human body when it comes to weight? In other words, if your body goes above or below your set point range, does your metabolism rev up or slow down to get your weight back in your zone? The metabolic set point theory was originally developed by William Bennett and Joel Gurin to explain why repeated dieting is often unsuccessful in producing long-term change in body weight or shape.[58] The theory was first described in the early 1980s, and since then numerous scientific studies have tried to find evidence for, and document, a set point that regulates human body weight.[59] But as so many researchers have found, multiple variables factor into the weight equation; while there is some evidence to support the idea that an internal active control of body weight exists and has a sort of set point, it is not a fixed or immutable set point such as body temperature.

Not only does the body have more than one way to defend its fat stores, but body weight is the product of three main forces: (1)

genetic effects (inherited DNA); (2) epigenetic effects (heritable traits that do not involve changes to DNA); and (3) the environment (diet, exercise, chemical exposure, sleep, and so forth). Regulation of body weight is complex and dynamic. It may even be asymmetrical—more effective in response to weight loss than to weight gain, which could help explain why it is easier to gain weight and keep the weight on than lose it permanently. We have learned a lot and have much more yet to discover, but at least we are beginning to understand the secret life of fat cells and how their behavior can change everything when it comes to weight.

Once you reach adulthood, the total number of cells in your body does not change much. Each type of cell has an average life span (for example, sperm cells have a life span of about three days; colon cells last only about four days; skin cells die and slough off after about two or three weeks; whereas brain cells typically last an entire lifetime). The average fat cell (known as an adipocyte) lives for about ten years.[60] Although we know how the body creates new fat cells, we know little about how this number is determined. Early life programming tells the body how many adipocytes it is supposed to have, and the body will defend that number. If you remove some fat cells by liposuction, the body will restore the number of cells, but not necessarily in the same place as they were.[61,62] Recent studies show that your diet can cause the number of fat cells to increase, but we do not yet know how to reduce the number of fat cells permanently. In order to significantly shrink the size of existing fat cells, you need to make substantial changes in what and how much you eat and maintain these changes. If you have increased your baseline number of fat cells through prolonged poor eating habits or from the impact of obesogens, you will struggle more with managing your weight because those fat cells are programmed to contain at least a minimum amount of fat. And as we will learn in chapter 3, small fat cells produce the least amount of the satiety hormone, leptin.

Later on, we will see how obesogens can affect one's metabolic set point. But for now, let's stick to the basics of normal fat metabolism and the road to obesity.

THE FAT FACTS

Contrary to popular belief, the road to obesity is not a straightaway. The human body is a lot more sophisticated and complex than most people realize or appreciate. As we further our understanding of fat metabolism and the development of obesity, we uncover fascinating new facts about fat. Fat cells are not all created equal. Moreover, different fat depots have disparate functional characteristics in terms of how easily they build and burn fat.

When I was in graduate school, the general thinking was that fat cells were mainly storage bins for excess calories. In other words, fat cells were seen as passive cellular containers for stored energy. They stockpiled energy and released it when needed. But it turns out that was a patently myopic perspective. Our understanding of fat tissue biology has progressed rapidly; fat has been recognized as a bona fide endocrine organ since the 1994 discovery of the satiety hormone, leptin,[63] and the master regulator of fat cell development, peroxisome proliferator-activated receptor gamma (PPARγ).[64] Fat tissue is an active endocrine organ that plays a key role in human physiology. Fat tissue releases more than twenty hormones, some of which are related to appetite and metabolism, that can communicate with other tissues, such as the brain, liver, pancreas, and immune system. Fat tissue, to be sure, is a key regulator of energy and nutritional balance in your body, and it is anything but passive.

The old adage in real estate circles is that the three most important factors in the value of a property are location, location, and location. This is also true for fat: where your body caches fat has health consequences, particularly with respect to heart and artery (cardiovascular) disease, type 2 diabetes, and stroke. As with cholesterol, there is "good fat" and "bad fat." Fat stored inside your abdominal cavity (around your organs) is associated with most of the health risks of obesity and is considered to be "bad fat." In contrast, fat stored under your skin (subcutaneous fat), like the kind of fat you see under your arms or as rolls on top of your abdominal muscles, is not associated with these health risks, although you might consider it the most unsightly.

To fully comprehend fat metabolism, it helps to first have a general understanding of how hormones work. Endocrine hormones are biological messengers produced in glands such as the pituitary, adrenal, ovaries, and testes that travel through the blood to other parts of the body, where they exert their effects. These biological messengers typically act at vanishingly small concentrations and have many important jobs in the body. Hormones act via highly specific hormone receptors. The presence or absence of hormone receptors determines whether a particular cell or tissue will respond to the hormonal signal or will instead ignore it. Broadly speaking, there are two types of hormone receptors. The first type are those that live on the cell surface and exert their function through a cascade of cellular messengers. These cellular messengers are called "second messengers" because they relay messages received at the cell surface from the first messenger—the hormone itself. For example, insulin signals through the insulin receptor, which acts through a group of second messengers that eventually elicit changes in gene expression. The second broad type of hormone receptor lives in the nucleus within the cell. It is what we call a nuclear hormone receptor, or nuclear receptor. These receptors are "ligand modulated transcription factors"; you can view them as molecular machines that bind to their specific hormone and directly regulate the expression of target genes. The estrogen and testosterone receptors are two examples of critical nuclear hormone receptors in the body.

There are more than fifty different hormones and related molecules that together regulate nearly every bodily process and are critical to the function of almost every tissue and organ. Hormones can regulate metabolism (how fast or slow it runs, whether you store or burn calories), growth and development between birth and maturity, tissue function, your moods, and much more. The diagram on the next page illustrates some of the key parts of the endocrine systems in the body and what hormones are produced by each.

At some point during your formal education you probably learned about the hypothalamus and pituitary glands in the brain, the sex glands (ovaries in women, testes in men), and the adrenals. Hormones are important for vertebrates and invertebrates and in many cases are very similar across species. The most well-known hormones are the ones we deal with daily, such as estrogen and progesterone, which regulate the monthly cycles of the female reproductive tract; testosterone, which gives men muscle and, in part, controls aggressive behavior;

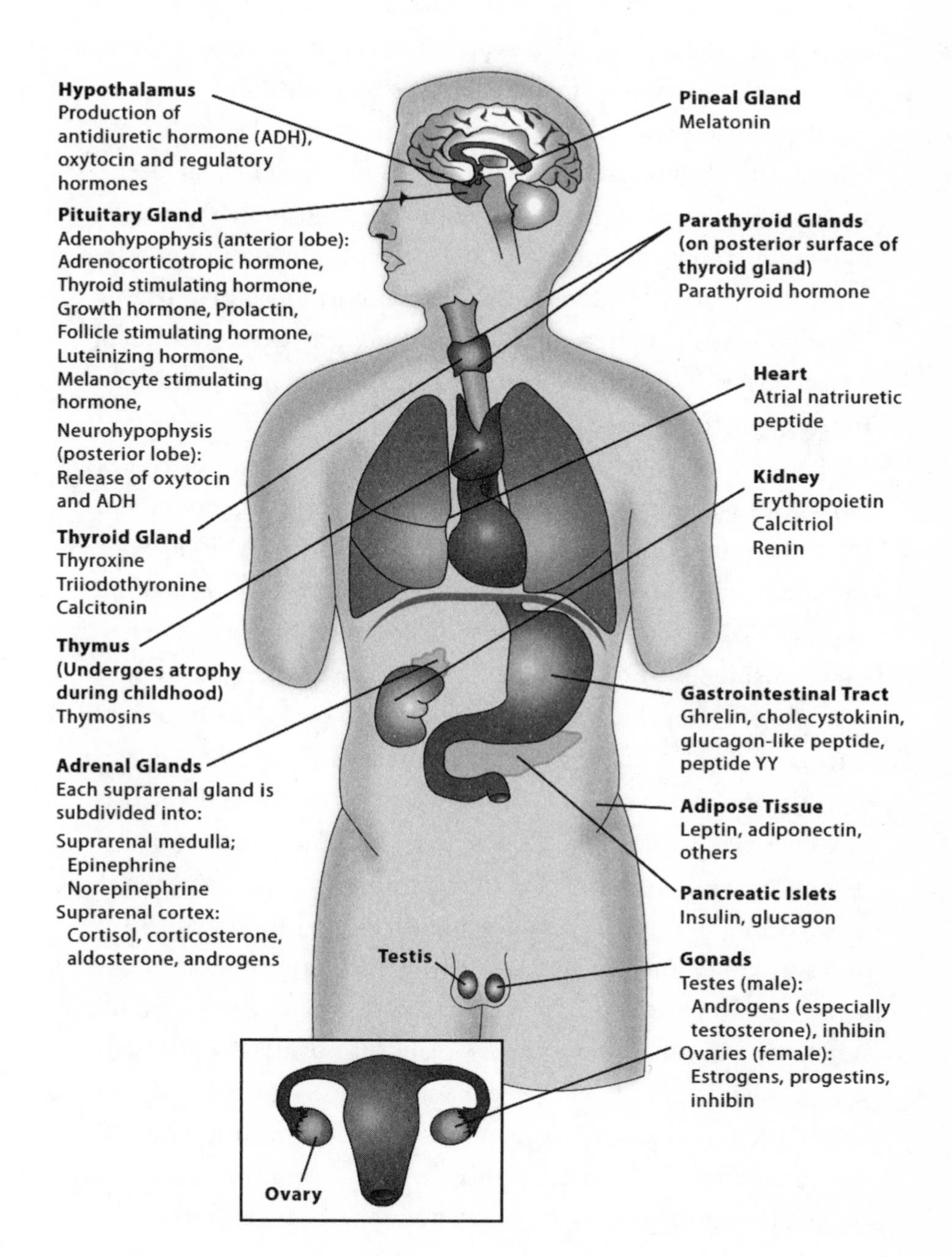
Hypothalamus
Production of antidiuretic hormone (ADH), oxytocin and regulatory hormones
Pituitary Gland
Adenohypophysis (anterior lobe): Adrenocorticotropic hormone, Thyroid stimulating hormone, Growth hormone, Prolactin, Follicle stimulating hormone, Luteinizing hormone, Melanocyte stimulating hormone,
Neurohypophysis (posterior lobe): Release of oxytocin and ADH
Thyroid Gland
Thyroxine
Triiodothyronine
Calcitonin
Thymus
(Undergoes atrophy during childhood)
Thymosins
Adrenal Glands
Each suprarenal gland is subdivided into:
Suprarenal medulla;
Epinephrine
Norepinephrine
Suprarenal cortex:
Cortisol, corticosterone, aldosterone, androgens
Testis
Ovary
Pineal Gland
Melatonin
Parathyroid Glands
(on posterior surface of thyroid gland)
Parathyroid hormone
Heart
Atrial natriuretic peptide
Kidney
Erythropoietin
Calcitriol
Renin
Gastrointestinal Tract
Ghrelin, cholecystokinin, glucagon-like peptide, peptide YY
Adipose Tissue
Leptin, adiponectin, others
Pancreatic Islets
Insulin, glucagon
Gonads
Testes (male):
Androgens (especially testosterone), inhibin
Ovaries (female):
Estrogens, progestins, inhibin

glucocorticoids such as cortisol, which regulate our response to stress; and thyroid hormone, which controls how many calories our bodies burn at rest. Insulin controls how much glucose is in our blood; insufficient insulin can lead to diabetes and a host of other health complications. Melatonin is required for our sleep/wake cycles. I could go on about many more hormones and their numerous important functions, but this introduction should be sufficient to illustrate the key role that hormones play in our lives and why disrupting their function can lead to many adverse health effects. Now let's look more closely at the hormones controlling whether you get fat or stay lean.

The dominant female (estrogens) and male (androgens) sex hormones play important roles in determining the amount and location of body fat throughout much of our adult lives. Your age and hormonal stage in life (for instance, a twenty-something versus a sixty-something) largely determine how and when these hormones get produced. The ovaries make the highest amount of estrogens, mostly estradiol, in women of childbearing age. These compounds induce monthly ovulation during an active menstrual cycle. In fact, the word "estrogen" is defined as any chemical that makes the uterine lining proliferate (grow). Young women produce large amounts of estradiol in their ovaries, while young men produce high levels of androgens, such as testosterone, in their testes; levels of both hormones decrease as we age. In men, and in postmenopausal women, estrogens do not come principally from the sex glands. Instead, fat and a variety of other types of cells produce estrogens, which mostly act where they are made rather than being secreted in large amounts into the blood.[65] The natural shifts in these sex hormone levels during the aging process are associated with changes in body fat distribution.

Changes in growth hormone production as we age also affect body fat. Growth hormone helps stimulate cell growth and division, is pumped out by the pituitary, and is essential during our younger years when we are developing rapidly—growing taller and building bone and muscle. Levels of growth hormone naturally begin to wane as we age, and this slows down our metabolism. Growth hormone supplementation is becoming popular for its "antiaging powers" (that is, to spur weight loss, improve the appearance of skin, and increase muscle mass more easily), but such use is controversial and may be associated with a heightened risk for cancer and an earlier death.[66]

Simplistically, you might think that restoring hormone levels to what they were when we were young is a good idea; perhaps you have seen pictures of a certain seventy-something doctor who looks like a young bodybuilder in advertisements on the Internet. Searching for the hormonal "fountain of youth" is a big business; however, it should be viewed in this light and not as a pursuit of good health. What is good for young people is not necessarily good for older adults. In reality, the same hormones may have different effects across the life span and result in different outcomes on health and disease. This is an important concept to understand because it relates directly to how substances such as obesogens and EDCs can have different effects across the life span. As you are probably aware, after menopause women tend to gain weight. This shows that loss of estrogens promotes weight gain and, therefore, that estrogens prevent weight gain. However, as I will explain, estrogens play a different role earlier in life. Exposure to excess estrogens before birth or during early life promotes obesity later in life. So the way an estrogen-like substance affects a prepubescent girl will be different from how that same substance affects a grown woman. This will become very important in our discussion of endocrine disruption later in the book.

Fat in your body serves many functions beyond keeping you warm, insulated, and cushioned. But remember, location matters. You might want to melt away the unsightly fat under your arms and on your thighs, but you would probably be better off worrying about the fat you do not see because it is wrapped snugly around your internal organs. People who outwardly do not appear to be obese can have high levels of harmful but hidden visceral fat that puts them at a much higher risk for diseases than people who appear fatter because they carry their weight as rolls of the unsightly subcutaneous fat most of us want to be rid of.

What is it about visceral fat that strongly links it to disease risk? Although scientists are still working out all the details, there are plenty of clues pointing us toward some answers. But to understand these clues, you must first understand the different types of body fat and how location dictates their impact on the body.

TYPES OF BODY FAT

Broadly speaking, body fat can be split into two fundamental categories: essential fat and storage fat. Essential fat is necessary for normal, healthy functioning and is found in relatively small amounts in your bone marrow, organs, central nervous system, and muscles, among other places. Men do not need as much essential fat as women (about 3 percent versus 12 percent of body weight) because essential fat in women includes fat in their breasts, pelvis, hips, and thighs that is critical for normal reproductive function. It follows, then, that reducing your body fat to near or below these levels, no matter how much you like your new appearance, will have negative health consequences. This could explain why some elite female athletes lose their monthly periods and why underweight women often suffer from infertility. Storage fat, on the other hand, accumulates beneath your skin (subcutaneous fat), in your muscles, and in specific areas inside your body. Overall, women generally have a higher percentage of body fat than men for childbearing reasons; they need that extra fat to meet the demands of pregnancy and lactation.

There are other ways to categorize body fat. The above categories of fat depots relate mostly to white body fat. New research in the past few years has revealed other distinctly colored fat types—from brown to beige in addition to white.[67,68]. Each of these different types of fat comes with unique molecular properties and health implications.

Brown fat is thermogenic—it uncouples metabolism from the formation of energy, which causes heat to be produced for the body (that is, it "burns" calories to produce heat). For example, human infants have quite a bit of brown fat (up to 5 percent of their body mass), which functions to keep them warm. Unlike white fat, which stores energy and produces hormones that are secreted into the bloodstream, brown fat tissue has many blood vessels. Brown fat acts more like a muscle in its ability to burn triglycerides to produce heat. In fact, brown fat cells and muscle cells are derived from the same precursor cell type. Although it was originally thought that only babies had brown fat, in 2009 researchers found small amounts of brown fat in adults. It is thought that an adult of normal weight will carry about two to three ounces of brown fat.

Beige fat, which was just discovered in 2012,[69] is an additional type of "good" body fat, although we know less about this fat and how it works. Beige fat is another form of thermogenic fat that has properties of both brown and white fat. Beige fat develops in areas of white fat in response to various activators, such as exposure to cold. Relatively little is known about the origins of beige fat. One school of thought holds that beige fat is derived from white fat, whereas others have shown that beige fat is produced in response to cold from a type of stem cell located around blood vessels in fat.[70] The activities of both brown and beige fat cells reduce metabolic disease, including obesity, in mice and are associated with leanness in humans. Induction of beige fat is being explored as a promising new therapy for type 2 diabetes.

In distinct contrast to the relatively small amounts of brown and beige fat found in our bodies, white fat is the plentiful energy-storing type of fat found throughout, particularly in the hips, thighs, and belly and under the skin. White fat is intended to store energy in the form of triglycerides, and the most benign place for fat to be stored is subcutaneously. Excess visceral (internal) white fat storage is unhealthy; you will often see having too much of this fat referred to as abdominal, central, or android obesity (they all refer to the same thing). Visceral white fat is metabolically different from subcutaneous white fat. Visceral fat actively releases fatty acids,[i] inflammatory compounds, and hormones that can ultimately lead to higher low-density lipoprotein (LDL, or bad) cholesterol, elevated triglycerides, increased blood pressure, and insensitivity to insulin, which often leads to type 2 diabetes.

---

[i] Fatty acids are the building blocks of the fat in our bodies and in the foods we eat. When you digest food, your body breaks down fats into fatty acids, which can then be absorbed into the blood. Fatty acid molecules are usually joined together in groups of three to glycerol, forming a molecule called a triglyceride.

One long-established explanation for the toxicity of visceral fat has been its relationship to an overactive stress response in the body. Increased stress leads to higher production of glucocorticoids (such as cortisol), which raise blood pressure, elevate blood glucose levels, and increase the risk of cardiovascular disease. These are definitely important factors, but we also need to consider the role of the liver and the concept of lipotoxicity—a metabolic problem that results from the accumulation of fat molecules in tissues and organs where they do not belong.

Unlike other types of body fat, visceral fat cells are unique in that they release their metabolic products directly into the portal circulation—the system of veins that carries blood from your intestines to the liver through the portal vein. Consequently, visceral fat cells that are enlarged and full of excess triglycerides dump free fatty acids straight into the liver. Many of these are converted by the liver back into triglycerides, some of which are stored. However, some of these fatty acids are not easily bound to glycerol by the liver and are released into the general circulation. This allows them to collect in places that are not intended to store fat, such as the pancreas, heart, and other organs. This can lead to organ dysfunction and, in turn, dysregulation of insulin, blood sugar balance, and cholesterol. All of this activity fuels biological pathways that can result in increased and widespread inflammation, which is not a good thing for overall health.

Large fat cells leak fat that attracts immune cells such as macrophages, which try to pick up this excess fat. Leaky fat cells also produce proteins that induce local inflammation, attracting other immune cells, leading to sustained inflammation. Inflammation per se is not necessarily destructive; inflammation is an important first step in healing from an injury. Inflammation is necessary for survival and serves as an indication that the body is trying to defend itself against something likely to be harmful. Without inflammation, we would not be able to combat foreign invaders such as pathogenic bacteria, viruses, toxins (natural poisons), and toxicants (man-made poisons). Inflammation kicks off the natural healing mechanisms in our bodies, temporarily revving up the immune system to take care of, say, a sprained ankle or cold virus. Inflammation is intended to be a temporary response that stops when the healing process is under way.

But what happens when the process is always "on" and the immune system is permanently keyed up? The biological substances (hormones, lymphokines, free radicals, prostaglandins, nitric oxide) produced during the inflammatory process harm cells throughout the body when they are continuously present. This type of inflammation is systemic—a slow-boil full-body disturbance that is usually not confined to one particular area. The bloodstream allows it to spread throughout the body; hence, we have the ability to detect this kind of widespread inflammation through blood tests. A classic marker that your doctor may measure in blood tests is C-reactive protein (CRP), which indicates an ongoing inflammatory process. One of the most important discoveries in modern medicine has been recognition that chronic (systemic) inflammation underlies many, if not most, chronic illnesses, including obesity.

Levels of inflammation may, in fact, help explain the difference between people who are obese yet apparently remain metabolically healthy and people who are obese and metabolically unhealthy. In general, obesity is linked to a higher risk of diabetes and heart disease, but some obese people do not develop high blood pressure and unhealthy cholesterol levels—factors that increase the risk of metabolic disorders. This phenomenon makes up as much as 35 percent of the obese population. A 2013 study showed that metabolically healthy people—both obese and non-obese—had lower levels of several markers of inflammation, such as CRP, in their blood. No matter what their body mass index was, people with favorable inflammatory profiles also tended to have healthy metabolic profiles.[71]

Study after study shows that there is a strong relationship between visceral fat and high levels of inflammation. In addition to visceral fat powering up inflammation, it can become inflamed itself, thereby generating a vicious cycle. Researchers from the Mayo Clinic showed in 2015 just how bad excess abdominal fat can be, irrespective of overall weight. Analysis of fourteen years' worth of follow-up data collected from 15,184 people who participated in the NHANES III survey (Third National Health and Nutrition Examination Survey) found that people considered to be "normal" weight but who had flabby midsections (central obesity) were twice as likely to die from heart disease as those considered obese but whose fat was distributed throughout their bodies.[72]

These findings further point to the adverse effects visceral fat has on the body compared with subcutaneous fat. We also know that the increased mortality risk accompanying higher ratios of visceral fat is likely due, at least in part, to increased insulin resistance. When you are insulin resistant, your cells do not respond to the hormone insulin, which is key to their ability to use glucose from the blood for fuel. In turn, this causes the insulin-producing beta cells of the pancreas to synthesize and secrete more insulin. Insulin resistance almost always leads to type 2 diabetes because the beta cells will eventually become incapable of producing enough insulin to meet the demand. Diabetics have high blood glucose levels because their bodies cannot transport glucose into cells, where it can be safely stored for energy. Instead, this excess blood glucose has pathological effects on the pancreatic beta cells and on vascular endothelial cells (the cells lining blood vessels). Most of the negative consequences of type 2 diabetes involve defects in circulation, particularly in the small vessels in the eyes and kidneys and near nerves. About 80 percent of type 2 diabetics are overweight or obese, establishing a strong link between fat and diabetes.

THE BIRTH OF A FAT CELL

In humans, fat tissue can first be observed by the fourteenth week of gestation, the time when the fetus is about the size of a lemon. This is followed by a second period of increased fat cell proliferation that continues after birth and lasts through adolescence. Fat cells are replaced at a rate of about 10 percent per year in adulthood; thus the tissue is not static. The contents of a fat cell change over time, and fat (and chemicals contained in this fat) stored in a cell that is recycled must necessarily be liberated in the process. Fat cells are derived from mesenchymal stem cells (MSCs), which are a special type of precursor cell found in the bone marrow and around blood vessels in fat tissue that can differentiate into a variety of cell types, including osteoblasts (bone cells), chondrocytes (cartilage cells), myocytes (muscle cells), and adipocytes (fat cells), depending on the signals they receive from surrounding cells and tissue. Once formed, fat tissue, as you know by now, is dynamic. Fat cells are capable of growing to fifteen times their original size.[73]

We don't know exactly what triggers the body to generate more fat cells and to maintain its set number of fat cells. It is recognized that by about the end of adolescence, our bodies "know" how many fat cells they will have, but how this happens remains a mystery. It was previously believed that once the number of fat cells is set, this number does not change and that these cells last for a lifetime. In a clever 2009 study that changed this thinking, Kirsty Spalding and Erik Arner, working at the Karolinska Institute in Sweden, used a by-product of atmospheric nuclear testing (radioactive carbon, or ^{14}C) to show that fat cells are actually continually replenished throughout life. The average fat cell lives about ten years, and around 8.4 percent of human white adipocytes turn over every year.[60] What still remains largely unknown is how the body knows the correct number of cells to regenerate. Developmental biology is providing some answers about the developmental origins of fat cells as well as information about what types of signals trigger the formation of new fat cells from local stem cells.[73] One thing is for sure, no matter what underlies the process: more fat cells—and bigger fat cells—means more weight.

HOW CALORIES LEAD TO WEIGHT GAIN

Have you ever wondered what happens after eating a meal? How does food become part of your fat if you overindulge? It is important to understand how this process works because it sheds light on how erroneous many of the generalizations made by fad diets are. Here is a very simplified overview.

Digestion starts in your mouth. Saliva contains enzymes that break down starches in the food to simple sugars. The food then moves to the stomach along with any fat and water. Once there, the food gets churned up and transformed with the help of enzymes such as pepsin, which digests protein, and hydrochloric acid. These chemicals further break down the food and turn it into a mixture of gastric juices and partially digested food called chyme, which then enters the first part of the small intestine called the duodenum. This is where bile, secreted from the liver and stored in the gallbladder, mixes with the chyme to emulsify the fat, permitting it to be digested. A variety of enzymes made in the pancreas also join the party to further break down the carbohydrates, fat, and protein. This process ultimately thins out the mixture into a fluid form so nutrients can be more easily absorbed through the lining of the small intestine. Here the paths of fats, carbohydrates, and proteins diverge. Fats and fat-soluble molecules travel readily across the cell membranes. Nutrients, proteins, and carbohydrates require specific transporter proteins that move them either actively or passively across the cell membrane. If the cell does not need to expend energy, the process is said to be passive, whereas transport that requires energy is active.

Once carbohydrates have been broken down into simple sugars such as glucose and fructose, they can be moved into the bloodstream for delivery to cells and tissues. Some carbohydrates become stored in the liver as glycogen, and whatever is left is converted to fat and cached in fat cells. The fats also go into the bloodstream, but they are destined for the liver, which burns some of the fat and converts some to other substances, such as cholesterol. The remainder is destined for storage in fat cells.

Proteins are broken down into smaller fragments known as peptides, which are further dismantled to become amino acids that can then be transported across the lining of the small intestine to enter the bloodstream for transport to cells. From here, some amino acids are used to build new proteins. Superfluous amino acids can end up in fat cells, albeit in a roundabout way. When protein is first metabolized into amino acids, some types of these amino acids can be converted into glucose, which your body also uses for energy, through a process called gluconeogenesis. If your cells have enough glucose, and glycogen stores in the muscles and liver are full, the excess glucose is converted into fat and stored. Other types of excess amino acids are converted by the liver into "ketone bodies," which are much beloved by Atkins and Paleo devotees. Ketone bodies can be used to produce energy in most cell types. So technically speaking, even extra protein can eventually become part of fat tissue. Thus, the commonly held idea that eating lots of protein and avoiding carbohydrates will only "put muscle on your bones" is false. Excess of any type of food will end up in fat cells.

I should also note that how your intestines absorb nutrients depends on the health and function of the trillions of bacterial cells that collaborate with your digestion and influence how many calories in the form of nutrients your body takes in. Depending on the exact types of bacteria in their intestines, some people absorb more calories than others and thus have a higher risk of overweight and obesity. Scientific research is also under way to understand the role of metabolites produced by the gut bacteria that may contribute to body weight and disease. Some types of these bacteria and the small molecules they produce (that also end up being absorbed into the circulation) have been implicated in obesity and diabetes, whereas others have protective benefits against weight gain and metabolic disorders. We will talk more about this in chapter 5.

THE INS AND OUTS OF A FAT CELL

To understand how fat cells fill up and empty out, it helps to know a little bit about lipoproteins. Fats (lipids) do not dissolve in water like other food items we consume. They must first be broken down to be absorbed and circulated. The fats you eat and the fats your liver can create from carbohydrates are bound to protein molecules for transport throughout the blood. In fact, all sorts of non-water-soluble molecules (fats, steroid hormones, some vitamins) must be bound to proteins for transport through the blood. Lipoproteins (lipids bound to proteins) are specific proteins that are recognized by receptors on the surface of the cells that are destined to process them. Different lipoproteins are charged with specific responsibilities depending on which cellular receptors they can bind to. For example, low-density lipoprotein (LDL) receptors on the surface of many types of cells recognize cholesterol-rich LDL from the blood so the cells can extract cholesterol. LDL transports lipids to the cells for use in important cellular processes such as making cell membranes, making the myelin sheath that surrounds nerves, and synthesizing cholesterol derivatives such as steroid hormones, bile acids, and vitamin D_3. LDL also transports excess fat to fat cells for storage, but some of it may end up in the liver and in arterial walls, where it gets stuck in plaque-forming foam cells that can eventually cause blocking of the arteries. In contrast, high-density lipoproteins (HDL) are recognized by HDL receptors on the liver cells that are responsible for recycling "used" cholesterol and removing fat from the body. The amount of HDL a person has is inversely correlated with risk of heart attack—that is, more HDL is associated with a decreased risk of heart disease.

Although fat cells are storage depots for fat, their "doors" cannot accommodate large molecules. Triglycerides are too big to easily pass through the membrane of a fat cell without breaking up a little bit first into smaller pieces. The individual fatty acids that make up the triglyceride must be released and then travel across the cell membrane. The major enzymes responsible for stripping down triglycerides into their component fatty acids and glycerol are called lipases. One of the most important enzymes in allowing fat to enter fat cells is lipoprotein lipase, or LPL. The glycerol molecule remains outside.

Fat cells do more than just shelter fatty acids supplied from outside. With the help of glucose from the blood, they also build their own triglycerides. So fat storage involves two key ingredients: LPL to liberate free fatty acids from LDL, and glucose to generate triglycerides inside the cell from these fatty acids and glycerol. All of this activity is coordinated to a large extent by insulin, which regulates how much glucose gets from the bloodstream into muscle, fat, and liver cells. More insulin, either from high sugar intake or from being insulin resistant, means that there will be more LPL and more glucose to transport into the fat cell, ultimately producing more fat storage.

LPL is not exclusively bad; it does not exist solely to load up fat cells. For example, the LDL receptor will move fat into a muscle cell, where it can be chopped up by LPL and burned for fuel. LPL helps to haul fat from the bloodstream into whatever cell happens to have an LDL receptor. In the absence of insulin, an enzyme called hormone-sensitive lipase, or HSL, conducts business in the other direction. HSL frees fatty acids from stored triglycerides so they can go into circulation and be used for energy. This is how your fat gets "burned." When there is less insulin around and more HSL activity, fat is more easily tapped for fuel. But when insulin levels are high, LPL takes control and moves fat into storage.[74]

This well-defined series of events by which these two metabolic processes balance each other out helps explain why we find that overconsuming carbohydrates precipitates much of obesity. When we eat carbs, that rapidly raise blood sugar, such as simple, refined carbs insulin levels also rise rapidly, especially when this activity is not countered with physical activity to burn the glucose. This creates the conditions for fat storage and fat retention by the processes described earlier. High insulin levels favor fat storage by inducing the expression of LPL. About the worst thing you can do for your body fat is to consume high levels of fat and simple sugars together. Unfortunately, this is known as "the Western Diet," the predominant diet in the United States (also referred to as the Standard American Diet or, appropriately, SAD).

While we all know that eating a large excess of anything—fat, protein, or carbs—can lead to obesity, we often forget that the body is in many ways self-regulating. It has numerous automatic feedback mechanisms to maintain balance ("homeostasis"). We have evolved to be physiologically resilient over a wide range of potentially harmful conditions. Our bodies are designed to store enough energy to survive between meals and even during periods of starvation without compromising vital organ systems or our physical performance. These feedback mechanisms involve not just the metabolic system, but also the endocrine, nervous, and digestive systems. I have hardly mentioned the role of the brain in all this, but it is also part of the picture. Without the brain, your body could not produce many hormonal signals at all, much less listen to the messages and perform certain functions, from the simple, such as overseeing digestion to the complex, such as coordinating communications, so you know when you are full.

GLITCHES IN THE SYSTEM

My reason for emphasizing the complexity of the human body is to underscore the point that obesity is rarely a condition that is "natural" to develop if you possess a healthy body whose functions are all working properly. To become obese typically requires enough glitches somewhere in this elaborate system to perpetuate circumstances that favor unhealthy fat cell proliferation. For example, a small percentage of people in the world are born with a genetic defect that prevents them from producing enough of the satiety hormone leptin, which is made by white fat cells and signals to the brain that you have sufficient fat stores, so stop eating. Luckily, these individuals can be treated with leptin to reverse their rare condition. But it is becoming common today that persistently high insulin levels coupled with low sensitivity to the leptin not only favor weight gain, but also disrupt our metabolic health. We become more prone to conditions such as type 2 diabetes, non-alcoholic fatty liver disease, high blood pressure, and abnormal cholesterol levels.

Carbohydrates are now considered the main culprit when it comes to obesity, and for good reason. Not only do they abound in refined form in our modern diets, but many trigger dramatic increases in insulin levels. This makes lots of glucose available for triglyceride storage. In a well-balanced system, insulin should simultaneously work toward suppressing appetite and move glucose into cells, but excessively high insulin levels move too much glucose into the cells. The brain then detects low levels of glucose in the blood and takes action to stimulate eating. It is as if the brain is tricked by the effects of the insulin spike and may think starvation is imminent. The brain needs a certain level of blood glucose to ensure its own proper functioning, so it activates signals that urge you to eat more, leading you to gravitate toward those carbs and perpetuate a vicious cycle. (Gary Taubes describes this unfortunate sequence of events in detail, and the basics of fat metabolism, in his book *Why We Get Fat*.[75]) Eating a balance of macronutrients—complex carbohydrates, healthy fats, and proteins—is ideal, as it promotes metabolic homeostasis. How so? It prevents dangerous insulin surges that create an imbalance in bodily hormone regulation. Proper energy levels in the blood can be maintained, and your preprogrammed system for controlling appetite and managing hunger can operate normally.

But we all know that many people do, in fact, rather easily become obese. Something has gone wrong with the system. If body fat is expertly regulated, then the obesity epidemic cannot be explained entirely by lack of restraint, gluttony, and sloth. Indeed, the obesity disorder is about abnormal growth of fat tissue, but the story rarely includes a look at the hormones and enzymes that manage that growth to begin with. And rarely does it consider the impact of obesogens.

THE OBESOGEN GLITCH

In early 2003, I was at a meeting in Matsuyama, Japan, that was about endocrine disrupting chemicals (EDCs), a subject area in which I was increasingly working that we will discuss more shortly. Here I heard a talk by Professor Shinsuke Tanabe from Ehime University about tributyltin (TBT), a compound used in antifouling paints to prevent invertebrate organisms from growing on ship hulls. My lab was collaborating with my Japanese friend and colleague Professor Taisen Iguchi to determine whether twenty priority EDCs of interest to the Japanese government, including TBT, could activate a nuclear hormone receptor I discovered when I was at the Salk Institute that we called the steroid and xenobiotic receptor (SXR). One major function of SXR is to regulate expression of enzymes that break down drugs and toxic chemicals (aka the xenobiotic response). Professor Tanabe's presentation described how TBT could turn female fish into males, and I wondered what exactly TBT was up to. The most obvious way to cause such sex reversal was to alter the function of estrogen or androgen signaling pathways.

Curious about what receptor TBT might be targeting, I asked my team back home in California to test TBT on our entire collection of nuclear hormone receptors to determine which one(s) it activated. To our surprise, instead of activating or inhibiting a sex hormone receptor, we found that TBT and related chemicals activated PPARγ and its cellular partner, the retinoid X receptor (RXR). This receptor dimer is the master regulator of fat cell development.[76] My team went on to show that TBT can spur fat cell precursors to become fat cells in vitro, that frog embryos exposed to TBT have their testes replaced by fat, and that mice exposed to TBT in utero have greater fat stores as adults. Six years later, we documented the most stunning finding of all: the offspring of exposed animals were also prone to store more fat. This proved how heritable chemical-induced obesity could be.

CHAPTER 3

The Science of Obesogens

Why Chemicals—Not Willpower—Could Be Holding the Remote Control to Your Weight

We can now prove that environmental influences will trump genetics when it comes to making us fat as a result of the effects that obesogens have on physiology and, especially, fat metabolism. Obesogens can create permanent glitches in a system that is otherwise expertly regulated. Obesogen exposure programs the body to store more of the calories you consume as fat and mobilize less of your stored fat for burning.[77] In this chapter, I will complete the story with the details of exactly how this happens. Since my laboratory introduced the concept of obesogens to the scientific community with the discovery that tributyltin exposure makes mice fat, numerous other chemicals have been identified as obesogens or potential obesogens. For the record, I define a bona fide obesogen as a chemical known to increase weight in a living organism, whereas a ***potential*** obesogen can induce cultured cells (grown in a laboratory dish) to become fat cells or control gene expression pathways known to promote the development of fat cells and obesity.

Let's get up close and personal with how obesogens work in the body, the way they reprogram cell fates and adversely change physiology, and how their effects can be passed on to future generations. There are quite a few known obesogens that can act both directly and indirectly through a variety of mechanisms.

I will be discussing some basic biology and endocrinology, subjects that are key to understanding obesogens in their complexity. You will soon know much more about how people get fat in the twenty-first century and what must be done to combat it.

HORMONE HAVOC: ENDOCRINE DISRUPTING CHEMICALS

Many obesogens belong to a larger family of troublesome compounds called endocrine disrupting chemicals, introduced earlier. So let's go there first. The field of endocrine disruption is historically rooted in reproductive endocrinology and wildlife biology. The idea that synthetic chemicals could change hormonal systems to cause adverse effects in wildlife and humans was first proposed in 1991 at the watershed Wingspread Conference held in Racine, Wisconsin, organized by Theo Colborn. In 1996, Theo and Pete Myers sounded the alarm and conveyed these concepts to the public in their book, *Our Stolen Future*.[78] In a nutshell, Theo and Pete identified particular negative effects in animal populations from a number of studies around the world that could be explained by inappropriate changes to their endocrine systems caused by chemical exposure. In my world, EDCs are defined as chemicals that come from outside the body (including pharmaceuticals), or mixtures of such chemicals, that interfere with any aspect of hormone action.[79] EDCs mimic or interfere with the actions of natural hormones produced in our bodies, leading to disrupted physiology and a variety of harmful effects. This definition differs from that used by the U.S. Environmental Protection Agency (EPA) and the toxicology community, who would add, "And cause adverse effects in living organisms."[80] To an endocrinologist, disruption of hormone action is adverse in itself. We will talk more about EDCs and how we are exposed later in the book. For now, suffice it to say that EDCs are ubiquitous and we are exposed to them in many, if not most, man-made products we encounter in our daily lives.

A very important concept to grasp when it comes to the actions of hormones and how they impact physiology is that the same hormone at the same amount will have different effects on adults from those it will have on developing embryos, fetuses, or children. "Activational" effects are rapid, temporary effects that come and go with the presence and absence of the hormone. "Organizational" effects, on the other hand, are permanent because these change the structure and function of an organism in a way that cannot be reversed. Let's explore these two types of effects with examples and then bring them into context with EDCs.

Activational types of hormone action are important for physiological homeostasis—the process by which the body stays balanced. The hormone-receptor complex functions much like a thermostat. When hormones are present and bind to their receptors, physiological processes are triggered until the hormone is removed (or when the temperature is lower than the set point, the thermostat turns on the heat until the set temperature is reached, then it turns the heater off). Many hormones induce the expression of inhibitors of their own action so that we are not subject to runaway signaling, as if the thermostat were broken and the heater or air conditioner continued to run without stopping. When you are suddenly threatened, for example, adrenaline will rush through your body creating the prototypical "fight or flight" response to the perceived stress. When the threat passes, hormone levels wane and your body returns to normal. This is essential for overall homeostasis in the body. When all hormones and related molecules are balanced, the body works as it should: organs function properly, and your metabolism operates smoothly. Conversely, prolonged, significant variations in hormone levels can wreak havoc on bodily functions, especially those that control metabolism.

In contrast to activational effects, hormones can also act during fetal development and throughout early life and puberty until adult maturity. These "organizational" effects of hormone action permanently alter the organization, proliferation, differentiation, and size of cells, tissues, and organs. My colleague R. Thomas (Tom) Zoeller from the University of Massachusetts uses a vivid analogy to illustrate the difference between activational and organizational effects. If you expose an adult woman to sufficient levels of anabolic steroids such as testosterone, she will grow bigger muscles and more body hair, both of which will eventually disappear after the hormone is removed—this is activational. On the other hand, if you expose a female ***fetus*** to high doses of testosterone, her clitoris will instead develop into a penis. This penis will not turn back into a clitoris when the testosterone is removed—it is there permanently, an organizational effect. Organizational effects of hormone action explain how early life exposure to EDCs can lead to permanent effects on the exposed individuals. For example, they are the reason a baby's exposure to pesticides can permanently change his or her body—from altering metabolism to increasing the risk for cancers and other diseases—for life.

Rachel Carson first made the connection between synthetic chemicals and cancer in her classic book, *Silent Spring.*[81] By the time the Wingspread Conference was held almost thirty years later, it was widely understood that several pesticides were linked with cancers. Theo Colborn had a different and more expansive view. As a wildlife biologist, she noticed that top predators (birds, fish, mammals) in the Great Lakes region had a variety of reproductive defects that were also observed in their offspring. She worried that the ability of these species to reproduce was being compromised by their chemical exposures. While cancer is definitely frightening to most of us, impairing the ability of a species to reproduce is the surest way to cause its extinction—a far scarier prospect, I think.

Some of the most widely studied EDCs are chemicals that alter the balance of sex hormones in wildlife and contribute to adverse reproductive outcomes such as sex reversal and/or sterility in aquatic animals, including fish and marine animals. EDCs were first identified as substances that interfere with the action of estrogens, androgens, and thyroid hormone. These are in fact the primary focus of the Endocrine Disruptor Screening Program, run by the EPA, which was intended to test chemicals for their potential to adversely change hormonal signaling and protect the public from the harmful effects of endocrine disruptors. We know now that EDCs can alter hormonal signaling systems by tinkering with many more receptors than just these three hormones. In principle, any of the many types of hormone receptors previously discussed can be inappropriately regulated by EDCs.

But how do EDCs disrupt hormonal function? One obvious

way is for them to mimic naturally occurring hormones in the body, as DDT does with estradiol (the major estrogen in human females). DDT is a pesticide that was used widely in the United States before it was banned in the 1970s because it adversely affected wild bird populations and could potentially harm human health; today it is among the most well-known EDCs. Another way would be for EDCs to block activation of hormone receptors so that the natural hormone no longer works, as the major breakdown product of DDT, called DDE, does with the androgen (testosterone) receptor. EDCs can make cells more sensitive to stimulation by existing hormones or disturb normal hormone levels by inhibiting or stimulating the production and metabolism of hormones or by changing the way hormones are transported to target tissues.

Contrary to what the chemical industry wants us to believe, the effects of EDCs, just like those of hormones, can occur at very low levels. This is a key point, because you will often hear that "the dose makes the poison" and that the doses of EDCs and other toxic chemicals to which we are exposed are far too low to

harm us or to stimulate hormonal pathways in our body. Some industry apologists suggest that EDCs probably don't even get into our bodies when we are exposed to them as intended by the manufacturers.

If you are leaning toward believing this argument, I recommend that you read a book by Rick Smith and Bruce Lourie called *Slow Death by Rubber Duck*.[82] Smith and Lourie used themselves as experimental subjects and tested the hypothesis that no significant amounts of chemicals gets into our bodies from using products as intended. They established personal baselines by measuring concentrations of chemicals such as phthalates and PFOA in their blood. Next they did normal things such as spray an air freshener in a room or stain repellent on a couch, then sat in the room or on the couch and watched television. After a day or two they had their blood measured again and, surprise, surprise, significant levels of phthalates and PFOA were found. These are among the most ubiquitous of obesogens.

"The dose makes the poison" happens to be the central dogma of toxicology (the study of poisons). Traditional "dose makes the poison" toxicologists (that is, most government and industry toxicologists) assert that all substances (even water and air) are toxic and differ only in how poisonous they are (in other words, the amount it takes to kill you—their "potency"). This preposterous idea dates back to the so-called father of toxicology, the Swiss alchemist, astrologer, mystic, occultist, and physician Philippus Aureolus Theophrastus Bombastus von Hohenheim, more well-known as Paracelsus, who worked in the early 1500s. Paracelsus believed that poisons came from the stars and that "all things are poison, and nothing is without poison; only the dose permits something not to be poisonous."[83] In this view, everything is a poison above some threshold dose, below which no adverse effects occur.

Although dose is certainly important, the argument that the dose makes the poison is patently wrong in many ways and on at least three counts with respect to EDCs.[79] First, the endocrine system in our bodies is already active. Therefore, rather than an EDC needing to activate a system that is "off" (and might require some threshold amount to turn on), the hormonal system in living organisms that use hormones is already "on" and can easily be disrupted by small amounts of EDCs. Second, since EDCs can have effects at very low doses (less than parts per billion), the entire concept of a "safe dose" below which we can comfortably believe there will be no harm is completely false. "Safe doses" for both acute or chronic exposures are often established by the EPA at ***much*** higher than parts per billion. Third, it is impossible to formally demonstrate the existence of a threshold for EDCs; moreover, even if you assumed that much of the population is "safe" from the doses we experience, you can bet that there will be some segment of the population that ***is*** sensitive to the doses they experience. Each one of us carries our own susceptibilities to chemicals. But government regulators largely test only for overt poisoning and completely miss the boat on long-term effects, particularly of EDC exposure.

There are more than eighty-four thousand chemicals registered with the EPA in commerce, most of which are poorly studied, and more than eight million unique chemicals available for purchase worldwide. Current studies have identified approximately one thousand of these chemicals that meet the criteria of an EDC. The actual number of EDCs is likely to be much larger because there has been no systematic effort under way anywhere to identify which of this deluge of synthetic chemicals are EDCs.

As mentioned, these compounds are used in a wide range of consumer products, including food packaging, building materials, clothing and upholstery, personal care products, detergents and other cleaning agents, plastics, and medical equipment. They are also used as pesticides and in industrial processes, leading to unintended contamination of food, water, and air. This means that we can be exposed to EDCs through what we eat, breathe, and put on our skin. EDCs abound even in hospitals.

My point in bringing this up is not to give you a sense of hopelessness or doom but rather to emphasize that you are largely on your own with respect to protecting yourself from chemical exposure. Government risk assessors (officials who assess the potential risk of chemical exposure to health) are heavily influenced by the industries they regulate, and the situation will only become much worse as Trump-era appointees remove environmental laws and regulations one by one. As Tom Zoeller put it most eloquently, "Chemical risk assessment is a collaboration between government and industry to expose the public to toxic chemicals for profit." It is precisely that—how much of a potentially useful toxic chemical (for example, a pesticide or herbicide) can be put into the environment before people start dropping dead in the streets? So far, risk assessors are doing a reasonably good job at that—people are not dying from chemical poisoning in the streets. Beyond this, you are on your own—there is little testing for and protection from chronic effects of chemical exposure at the levels we experience every day.

Another big problem with EDCs is that many are persistent and can bioaccumulate in our tissues, meaning their concentration increases over time. They also can become biomagnified; that is, their concentration increases at successive levels of consumption from plants to animals to humans. Perhaps you know about mercury contamination in seafood. Mercury ends up in the oceans as a result of coal-burning power plants, mining, and plastic production, among other sources. It is taken up by algae, which are eaten by small animals, which are eaten by small fish, which are eaten by larger fish, and so on. Eventually, the top predators such as swordfish, sharks, bluefin tuna, tilefish, and others accumulate mercury levels that can be dangerous to humans. While some EDCs, such as BPA, are not thought to accumulate in the body, many EDCs can be stored in fat cells for years after exposure and passed on to children during pregnancy or when breastfeeding (and no, you can't do a cleanse or detox to remove them quickly). For this reason, in the late 1990s the World Health Organization (WHO) and United Nations Environment Programme (UNEP) adopted the Stockholm Convention on Persistent Organic Pollutants, which seeks to severely reduce the production and use of chemicals that do not degrade in the environment. Would you be surprised to learn that as of this moment, the U.S. Senate has not ratified this treaty, despite the fact that virtually every other country in the world has done so? Given the current political climate, it does not seem likely that we will ratify it anytime soon.

The chemical apologists have yet another argument in their arsenal that we hear frequently: no one has conclusively demonstrated endocrine disrupting effects in humans. The implication is that only controlled human trials such as those the U.S. Food and Drug Administration (FDA) requires to license drugs would provide persuasive evidence that a chemical causes harm to humans. Fortunately for people everywhere, it is unethical, immoral, and illegal to conduct such experiments on humans. Ironically, industry uses a selected subset of animal studies to support its claims that chemicals are safe, while demanding evidence from human studies, which would be illegal to perform, to refute these claims. This is a typical yet unsupportable double standard. In reality, only drug side effects, accidental human exposures, or occupational exposures can provide anything approaching cause-and-effect data for chemical exposure in humans.

Sadly, we have some examples. A pharmaceutical EDC called diethylstilbestrol (DES) was prescribed by obstetricians throughout the mid-twentieth century with the aim of helping women avoid pregnancy complications and miscarriage, despite evidence from animal studies that it caused cancer.[84] Regrettably, children born from DES-treated mothers ("DES sons" and "DES daughters," as they came to be known) were found to be at higher risk for certain cancers, infertility, miscarriage, and ectopic pregnancies, all of which had been observed in the earlier animal studies. This is another example of an organizational effect, because the children exposed during pregnancy suffered the permanent effects, whereas the mothers were relatively unaffected.

Accidental poisoning events are tragic, but they stand as important proofs of the effects of EDCs on humans. An industrial mishap in 1968 Japan led to the production of cooking oil containing polychlorinated biphenyls (PCBs) and related chemicals. Consumption of this contaminated oil, and foods cooked with it, caused Yushō disease, as it was called, led to the death of almost half a million birds, and sickened more than fourteen thousand people. The human effects were cognitive impairment in children, defects in the immune system, and irregular menstrual cycles (all symptoms of endocrine disruption), together with many other negative outcomes. An almost identical incident and effects occurred in Taiwan ten years later, which confirms that PCB exposure was the cause. Perhaps not surprising, similar effects were observed in animal studies and in wildlife exposed to PCBs. You might expect that we as a society (and particularly government regulators) should be smart enough by now to heed such lessons and prevent the harm that EDCs cause before exposures occur. Unfortunately, this is not the case. On the bright side, part 2 of this book will tell you how to avoid exposure to obesogens and EDCs of all types.

HORMONAL CONTROL OF FAT DEVELOPMENT

At about the same time that reproductive biologists and wildlife biologists became aware of EDCs, fat tissue was only beginning to become accepted as a bona fide endocrine organ itself, let alone an organ whose function could be disrupted by environmental chemicals.

The identification of fat as an endocrine organ was largely instigated by the discovery of leptin, which is one of a group of hormones that control hunger and fat storage. Leptin reduces the urge to eat by acting on specific receptors found in areas of the brain such as the hypothalamus, which regulates appetite (among other things). Many obese people are insensitive to leptin, which prevents them from responding to normal satiety signals. They continue to consume calories even though sufficient energy has been stored already (more on this shortly).

Another important finding clueing us in to the role of hormones in fat development has been the discovery of the master regulator of fat cell development, a nuclear hormone receptor with a long, terrible name that is an artifact of history: peroxisome proliferator-activated receptor gamma (PPARγ). In fact, PPARγ is a fatty acid receptor. Activating PPARγ initiates a program of gene expression controlling numerous genes involved in fat cell production, fatty acid synthesis, and storage. You will be hearing a lot about PPARγ because we have shown that it is a major target for chemicals that alter fat metabolism for the worse.

In addition to recognizing that fat tissue is an extremely active endocrine organ, we know now that it is highly connected to steroid hormones (estrogens, androgens, and glucocorticoids, the latter of which are a group of hormones involved with metabolism and the stress response system) and that fat maintains a close relationship with the immune system. It is becoming clear that disruption of fat tissue function could contribute to diseases beyond obesity alone. Excess or dysfunctional fat tissue can, in fact, have a hand in the development of diabetes, infertility, and even cancer. To really get a sense of how obesogens impact the body, it helps to understand their relationship with fat tissue.

FAT PROGRAMMING

Take a look at the two mice on page 79. They were raised in the same lab and given the same food and opportunity to exercise. The only difference between the two is that the mouse on the right was exposed to a tiny amount (5 parts per billion, the equivalent of about ten drops in an Olympic-sized swimming pool) of an obesogenic endocrine disrupting chemical (in this case, the synthetic estrogen diethylstilbestrol) for the first five days after birth. This brief exposure programmed the mouse to put on fat later—not until many months after the exposure stopped. Although there were no detected differences in caloric intake or energy expenditure throughout its life, it continued to fatten up.[85]

The notion that chemical exposures can turn your body into a fat-storage (as opposed to a fat-burning) machine, reprogram cells to become fat cells, and predispose you to become fat ***on a normal diet*** is daunting to ponder. How can this happen? In our own work with tributyltin (TBT), we showed that in mice exposed to very low levels of TBT in utero, mesenchymal stem cells (MSCs) found in the bone marrow and white adipose tissue are predisposed to become fat cells in far greater numbers than MSCs isolated from mice that were not exposed to TBT. MSCs are precursor/regenerative cells in our bodies responsible for first producing and later maintaining a large number of tissues, including bone, cartilage, muscle, fat cells, and some types of neurons. A key point to understand is that the developmental switch between fat and bone lineages is mutually exclusive: either the MSCs become fat cells or they become bone cells. Therefore, TBT exposure leads to more fat cells and fewer bone cells over time.

What we know so far is that one way in which TBT can induce weight gain is through that master regulator of fat cell development I mentioned earlier, PPARγ. The amino acid sequence of the PPARγ protein changes very little between humans and other mammals, and even between humans and more distantly related vertebrates such as frogs. PPARγ may be particularly susceptible to EDCs because it has a large "pocket" for binding to molecules and can accommodate many chemical structures. When a molecule capable of activating PPARγ enters the pocket, it forces PPARγ to change its shape, which then attracts a host of other cellular proteins that together bind to PPARγ-responsive genes and increase their expression. Among these PPARγ-responsive genes are many that are essential for fat cell development and function.

TBT promotes fat cell development in at least two ways. First, it induces expression of PPARγ in MSCs, which commits them to become fat cells. Second, it then activates PPARγ to turn on the genetic program that controls fat cell differentiation. My lab has also shown that when our animals are exposed in utero to TBT, but then never again, the damage has already been done: TBT causes a permanent effect on the metabolism of exposed animals, predisposing them to make more and bigger fat cells and to gain weight over time despite a normal diet. This was a heretical idea that met with considerable opposition from the medical establishment in the early days but has since been confirmed by multiple labs around the world working with TBT in different animals and with other obesogens in rodents. TBT can cause increased fat storage in fish, rats, and frogs. A group in Finland has even shown that levels of TBT in the placenta of pregnant women are closely correlated with weight gain in those babies when they are three months old.[86] These children are now twenty years old, and it will be very interesting to learn what effects maternal TBT exposure has had on their body fat as young adults if the researchers secure the needed funding for a follow-up study.

My lab has also shown that exposure to other chemicals can lead to weight gain in mice,[ii] such as the fungicide triflumizole, which is widely used on green leafy vegetables (one important reason to buy organic fruits and vegetables whenever you can).[87] Other labs have shown that exposure to estrogenic chemicals (for example, diethylstilbestrol, bisphenol A, and the pesticide DDT), organophosphate, organochlorine and neonicotinoid pesticides, flame retardants, alkylphenols, and phthalates all lead multiple animal species to increase fat storage.[7] Many of these are also linked with weight gain in human epidemiological studies. As we will see coming up, many of these chemicals are found in everyday products. You don't need to memorize a long list of chemical names. I will show you later in the book how to effortlessly steer clear of them without searching for them on a label (if there is one).

Many more chemicals have been shown in the lab to cause MSCs or other types of cells to differentiate into fat cells. These include alkylphenols, phthalates, flame retardants, and the plastics component bisphenol A diglycidyl ether, which is commonly used as a building block of epoxy resins and found in the lining of food and beverage containers.[23,88] A large number of these are agrochemicals (mostly fungicides)[89] that are sprayed on grains, fruits, and vegetables during conventional farming,

---

[ii] At the moment, the only obesogens that have a well-established mechanism of action are those that act through PPARγ, such as TBT, the fungicides triphenyltin and triflumizole, the antidiabetic drug rosiglitazone (also known as Avandia), and another fungicide, tolylfluanid, which acts via the glucocorticoid receptor. All of these chemicals lead to weight gain in animals. Avandia and a related drug called Actos cause weight gain in humans and serve as important proof of principle for PPARγ activators as obesogens in humans.

which further points to the health value of eating organic, as we will discuss in chapter 6. Turning cultured cells into fat cells does not prove that they will be obesogenic in humans or animals, but I do not know of a single chemical that caused cultured cells to become fat cells that did not also have the same effects when tested in animals.

TRANSGENERATIONAL EFFECTS OF EDCs AND OBESOGENS

One of the most startling findings about obesogens has been their power to be passed on to future generations. For example, Dr. Raquel Chamorro-García in my lab found that when she bred the offspring of TBT-exposed mice, the effects were inherited.[90] Raquel exposed pregnant mice to low levels of TBT (far below the levels at which adverse effects are observed in toxicological studies and comparable to allowable human exposures) in their drinking water throughout pregnancy. In genetic terminology we call these pregnant mice the F0 animals, and their offspring are labeled as the F1 group. The baby mice (pups) exposed while they were in utero got slightly fatter, had MSCs reprogrammed to make more fat cells, and also had fatty livers.

When Raquel bred these F1 mice to produce the next generation (F2), she found that they, too, showed the same effects, and this continued on for another two generations (F3 and F4).[77,90] Put in human terms, the exposed woman's children (F1), grandchildren (F2), great-grandchildren (F3), and great-great-grandchildren (F4) were all affected in the same way by her exposure to TBT during pregnancy. Although you might logically reason that TBT caused a permanent genetic mutation in the F1 generation that was then passed down to future generations, you would be wrong. We used so many different litters of pregnant F0 mice that there is no possibility the same mutation was induced randomly in all of them. Something else happened, and we will discuss that next.

EPIGENETIC—NOT GENETIC—CHANGES

So why did we even think to look for effects of TBT in the offspring of the exposed animals? The short answer is Professor Michael Skinner, a prominent reproductive biologist in the School of Biological Sciences at Washington State University. In 2005, Mike's group published a study that sent shock waves through biology that continue to reverberate today. Mike and his colleagues exposed pregnant rats to relatively high levels of a commonly used fungicide called vinclozolin (that happens to antagonize testosterone and is widely used on wine grapes, fruits and vegetables, and golf courses) or a pesticide called methoxychlor (that mimics estrogen and is now banned but still persists in the environment). In both cases they found that male offspring had low sperm counts as adults, together with prostate disease, kidney disease, immune system abnormalities, testis abnormalities, tumor development, and hypercholesterolemia (high cholesterol). When these males did succeed in impregnating a female, she gave birth to sons who also had fewer sperm and the same panoply of defects.[91]

Mike's group continued breeding the animals and found that the effects persisted until the F4 generation and beyond. This was the first demonstration that an environmental chemical had heritable, transgenerational effects. Because these defects intensified in the F3 group and its descendants, there is no possibility that genetic mutations were to blame. Instead, these effects were "epigenetic," which literally means "on top of genetics." Mike's hypothesis is that altered DNA methylation patterns—not genetic mutations in the DNA itself—were probably to blame.[92] Let me explain.

As we will discuss at length in the next chapter, DNA methylation is one way through which the expression of our genes is controlled without changing the DNA sequence. This is the essence of epigenetics. The changes in the epigenome are called "epigenetic marks" or "epimutations," and these are "read" by the transcription machinery in the cells, influencing whether the gene should be expressed or not. These epigenetic marks are important for our health and longevity and also how these traits are passed on to future generations.

While the concept of epigenetics is becoming well established, the new science studying the way that environmental factors influence gene expression is taking longer to be fully accepted because it harkens back to the two-hundred-year-old studies of Jean-Baptiste Lamarck. Lamarck believed that acquired characteristics could be inherited; he was wrong about the details of his theory (for example, that giraffes evolved from antelopes by stretching their necks and passing the longer necks to their offspring), but transmitting the effects of environmental exposures to our offspring is similar to Lamarck's theory about the inheritance of acquired traits. The opposite (dominant) theory—that inheritance is controlled entirely by our genes—is called genetic determinism.

Mike Skinner can tell many stories about his ongoing battles with the scientific orthodoxy to convince them that environmental epigenetics is a valid theory that should be incorporated into our thinking. This is always the fate of radical new theories: it takes quite a while before they are accepted by the scientific establishment. If you think about it, this is appropriate (although frustrating for the innovators) because we should not change long-standing, well-supported theories without strong, reproducible evidence. In the case of environmental epigenetics, that evidence is growing by the day.

At about the same time we were performing our transgenerational experiments with TBT, Mike and his group were screening a host of other environmental chemicals and pollutants for potential transgenerational effects. These included jet fuel, plastics ingredients, and more pesticides.[93,94] Once again, he noted that exposed animals had reproductive problems and quite a few other effects that were also passed down the generations. This time, however, they saw something new in their F3 animals: about 10 percent of the descendants of the F0 females injected with a mixture of BPA and phthalates were obese. This was not observed in the F1 and F2 generations, which were directly exposed to the chemicals, or in the control, unexposed animals. F3 animals were not exposed, which suggests that the transgenerational inheritance is mediated by epigenetic changes caused by the original exposure in the F0 animals that became apparent only when the effects of the direct exposure were gone. At least this is Mike's argument. These results were interesting, but not nearly as striking as the results Mike's team found when they tested DDT, the famous EDC pesticide. As was the case in the vinclozolin and methoxychlor experiments mentioned earlier, the F1 and F2 generations suffered a variety of abnormalities but were of normal weight. Remarkably, a full 50 percent of F3 animals—both male and female—were obese.[95]

Mike began to connect the dots in his head, considering both the marked rise in obesity rates among U.S. adults over the past few decades and pregnant women in the 1950s and 1960s who were exposed to DDT. There probably was not a woman who was pregnant in the 1950s or 1960s who was not exposed to DDT. Could the exposures in the 1950s have anything to do with the prevalence of obesity among adults today? We may never know for sure, because the kinds of experiments required to show this conclusively in humans are unethical to conduct (we cannot, nor do we want to, do double-blind, placebo-controlled clinical trials on humans using toxic chemicals), and the longest-running study of a human population that might demonstrate strong correlations between DDT exposure and obesity—the San Francisco Kaiser Permanente cohort—is now only in the F2 generation. But it is an intriguing observation nonetheless.

Although DDT is banned in the United States, it is still used in specific areas (for instance, in Africa) for controlling mosquitoes carrying pathogens such as malaria and dengue fever. So the possibility that DDT may cause transgenerational effects on obesity is a very real concern. The observation that multiple EDCs can cause transgenerational effects, including obesity, is something that needs to be addressed by regulatory agencies worldwide. Sadly, the wheels of government move very slowly indeed.

OBESOGENS AND NUCLEAR HORMONE RECEPTORS

As I briefly mentioned earlier, many endocrine disruptors act on nuclear hormone receptors to activate or inactivate them inappropriately. As we also discussed, EDCs that act as estrogen mimics can predispose animals to obesity. For instance, diethylstilbestrol (DES) is a potent estrogen that clearly exerts its effects through the estrogen receptors. Exposing a female fetus or newborn to DES will lead to massive weight gain later in life, but exposure to DES or other estrogens later in life leads to the opposite effect—leanness. Like DES, BPA also binds to estrogen receptors in the body, but we don't know whether it causes weight gain through estrogen receptors or some other receptors that it influences.

De-Kun Li, an epidemiologist and senior research scientist at Kaiser Permanente in Oakland, California, has also documented a relationship between BPA and obesity. He found that among school-age children in China, preteen girls (not boys, just as in mice) with higher BPA levels in their urine were more likely to be higher in weight, too.[96] In fact, after adjusting for potential confounders, a higher urine BPA level—at the level corresponding to the median urine BPA level in the U.S. population—was associated with more than double the risk of being at a weight greater than the 90th percentile among girls who were nine to twelve years old. Unfortunately, cross-sectional studies such as these, which measure chemical exposures and outcomes at a single time in life, cannot tell us whether the exposure preceded obesity. We need to follow people throughout their lifetime (longitudinal studies) to get closer to understanding causation.

INSULIN AND LEPTIN RESISTANCE: MORE WAYS THAT OBESOGENS CAN MAKE US FAT

To be sure, not all EDCs are obesogens, and not all obesogens are EDCs. For example, high-fructose corn syrup could be considered an obesogen as a result of its effects on metabolism and potential to increase body fat, but it is technically not an EDC. It does not disrupt the function of any nuclear hormone receptors, particularly the estrogen, androgen, and thyroid hormone receptors that the EPA associates with endocrine disruption.[iii] Similarly, the widely used agricultural weed killer atrazine, which contaminates water supplies throughout the Midwest, is an EDC but probably not an obesogen. It can negatively impact your hormones but as far as we know not in a way that will lead to weight gain.

It is important to understand that the biological mechanisms by which EDCs exert their effects are not mutually exclusive. EDCs can have direct effects on a particular target tissue by disrupting a specific receptor pathway, but they can also lead to widespread, sometimes subtle effects on multiple organ systems that ultimately promote obesity in the exposed individual and in subsequent generations. While we may have evidence that certain chemical obesogens cause animals to become obese or diabetic, we do not always know what biological pathways lead to this outcome. Research to date suggests that not all obesogens operate the same in the body even though the outcome—weight gain—is the same. We are fairly certain that TBT and triflumizole act through PPARγ, although they may also have other targets.

[iii] One could make the argument that HFCS is disrupting hormonal signaling in a different way that the EPA should be considering. Their view of endocrine disruption is too limited. Disrupting any of the many hundreds of hormonal signaling pathways (for example, leptin, insulin, growth hormone, and more) has adverse biological consequences that need to be accounted for. Perhaps even more important, disrupting any type of cellular signaling pathway, even if it is not a hormone-regulated pathway, can disrupt biological functions and lead to adverse consequences. My Japanese colleague Jun Kanno has introduced the concept of "signal toxicity"[97] to identify chemicals that disrupt any sort of receptor-based, cellular signaling pathway. This concept greatly broadens the field of endocrine disruption; therefore, the field needs to incorporate signal toxicity (or signaling disruptors) into the growing framework of metabolism disruptors.[7]

Although DDT is known to act like an estrogen, and DDE, the major breakdown product of DDT, counteracts the effects of male sex hormones such as testosterone (an antiandrogen), we are not sure how DDT promotes obesity in Mike Skinner's F3 rats. We don't know how many obesogens elicit heritable effects and whether these are carried by altered DNA methylation, histone methylation, or some other mechanism (we will come back to this in the next chapter). Whether a chemical can elicit permanent changes that can be passed on to the next generation of children and whether exposure occurs during a critical window of development (when germ cells are being programmed) can determine whether the effects of an obesogen will be temporary or permanent and transmitted throughout multiple generations.

In addition to reprogramming stem cells and encouraging the body to store more fat, EDCs could be promoting obesity in at least three other ways. One is by prompting cells to become insulin resistant, which makes the pancreas pump out more insulin to control blood sugar, leading to increased fat storage all over the body. Translation: You are more likely to turn the foods you eat into body fat. A second is by preventing the satiety hormone leptin from telling your body that you have had enough to eat. As mentioned, EDCs can make cells resistant to the leptin signal, thereby fueling weight gain. A third is that EDCs can inhibit the function of thermogenic brown fat. Although these possible mechanisms need to be better understood through future research, they are indeed credible. Let's start with the perils of insulin resistance.

If your body is no longer sensitive to insulin you will be on the road to experiencing serious metabolic issues, including diabetes, because insulin is the main hormone responsible for maintaining blood sugar balance. Weighing approximately 3.3

pounds in adults, the liver is the largest and most metabolically complex organ in the human body. Liver cells (hepatocytes) make up more than 80 percent of total liver mass and play a critical role in metabolism. The liver is the principal location of glucose storage as glycogen and the main source of glucose for all tissues of the body. (Recall from the previous chapter that glucose is a preferred source of energy.) Because the pancreatic veins drain into the portal venous system, every hormone secreted by the pancreas must traverse the liver before entering the circulation. The liver is a major target for pancreatic insulin and glucagon, the hormone that promotes the breakdown of glycogen to glucose in the liver. The liver is also where these hormones are removed from circulation and broken down. The body is about 40 percent by weight of skeletal muscle (30 to 40 percent in women, 40 to 50 percent in men), which is the major user of glucose in the body other than the brain.

One function of insulin is to stimulate storage of glucose as glycogen when glucose levels in the blood are high. When insulin levels in the blood go down, the body triggers glycogen breakdown so glucose can be released into the bloodstream for use as energy. If obesogens prompt the body to become insulin resistant, you may not be able to convert glucose to glycogen effectively, and glucose levels in the blood may rise to excessive levels. The pancreas responds by pumping out increasing levels of insulin, which promote fat storage and can trigger diabetes. If you are diabetic, by definition you have high blood sugar because your body cannot move glucose into cells, where it can be safely stored for energy. And if it remains in the blood, that excess sugar can damage most of the body, including kidneys, blood vessels, skin, cardiovascular system, and nervous system. Excess sugar also speeds cellular aging by binding to proteins, creating what are called advanced glycation end products (AGEs, for short), which are very damaging to cellular function.[98]

Keep in mind that insulin is a multitasker. When its levels are high and blood sugar cannot be managed well, it contributes to other biological processes. Insulin is an anabolic hormone that stimulates growth, promotes fat formation and retention, and encourages inflammation. When insulin levels are high, thyroid hormone, estrogen, progesterone, and testosterone homeostasis can also be destabilized. In turn, all of these imbalances have downstream negative effects on multiple body systems, including metabolism. These broader effects of excess insulin make restoring bodily balance (homeostasis) all the more difficult, even broken beyond repair.

Another way that EDCs may be acting to promote obesity is by causing leptin resistance, another co-conspirator in weight gain. Leptin is produced by your white fat cells and signals to leptin receptors in your hypothalamus, the part of the brain where your inner reptile lives. This ancient structure that predates humans (and even dinosaurs) is responsible for rhythmic activities (for instance, sleep-wake cycles) and a broad range of physiological functions in your body from hunger to sex. Broadly speaking, the more you can increase how sensitive your body is to leptin, the more normal your weight will be. By "sensitivity," I mean how leptin receptors in your brain recognize and use leptin to carry out various operations. When fat cells start to fill up and expand, they secrete leptin to tell your body that you have stored enough fat. Once the fat cells begin to shrink as their contents are burned for energy, the leptin faucet is slowly turned off. Eventually you are able to feel hunger again and the cycle starts all over. This is but one example of the many different mechanisms the body has to expertly manage energy metabolism in the name of survival. People with naturally low levels of leptin are prone to overeating. An important study published in 2004 showed that people with a 20 percent drop in leptin levels (due to sleep deprivation) experienced a 24 percent increase in hunger and appetite, driving them toward calorie-dense, high-carbohydrate foods, especially foods with a lot of sugar, starch, and salt.[99]

One more recently identified mechanism of obesogen action is to inhibit thermogenesis. Michele La Merrill and her colleagues at the University of California–Davis showed that prenatal and early postnatal exposure of rats to DDT led to reduced core body temperature in adulthood, together with decreased energy expenditure and intolerance to cold.[100] Intriguingly, the exposed animals developed signs of metabolic dysfunction when exposed to a high-carbohydrate diet that was similar to what is seen in humans with metabolic syndrome (prediabetes). These included glucose intolerance together with elevated levels of insulin and lipoproteins in the blood. This is the first demonstration that an obesogen can function by making the body essentially use less energy and not respond to normal environmental signals such as cold. I have no doubt that other obesogens will be identified that make us fat by reducing the "burning" of some of the calories we consume.

Here is a summary of how EDCs affect the body, ultimately preventing individuals from losing weight through usual means such as diet, exercise, and even fasting because their physiology has been reprogrammed. By mimicking or interfering with the actions of naturally occurring hormones in our bodies, endocrine disruptors that promote obesity can:

- Encourage the body to store fat and reprogram stem cells to become fat cells.
- Prompt the liver to become insulin resistant, which makes the pancreas pump out more insulin to control blood sugar, leading to increased fat storage all over the body. Translation: You are more likely to turn the foods you eat into body fat.
- Prevent leptin (the satiety hormone) from working properly in the body. Leptin normally signals the brain that you have had enough to eat. EDCs can make cells resistant to the leptin signal, thereby fueling weight gain.
- Prevent you from losing weight when fasting, which means that you may not lose weight effectively even when reducing caloric intake.
- Inhibit thermogenesis, the burning of fat to produce heat.

OBESOGENS AND METABOLIC SET POINTS

One area of my work looks at the potential consequences of obesogens on metabolic set points. As you may recall, a metabolic set point is the body's internal control mechanism that regulates metabolism to maintain a certain level of body fat that is regulated by the hypothalamus. We once thought that a person's metabolic set point was primarily genetically determined at birth and remained throughout life, but obesogens are changing the story.

Obese humans have more fat cells and probably developed them early in life by mechanisms I have been describing thus far in the book. We know that the minimum number of fat cells a person will have is programmed early in life. Fat cells develop beginning at the fourteenth week of human gestation, and the number of fat cells increases through adolescence. The number is largely fixed after that. If individuals are exposed to fat-inducing stimuli, such as obesogens, during this sensitive window of development, this can permanently increase their fat cell number[90] and thereby change their metabolic set point.[77] If this is true, as appears to be the case, the implications for weight gain are profound. The higher number of fat cells from the beginning of life cannot be reduced by diet, exercise, or even surgery—your body will defend this number of fat cells and add them back if you remove them, although not necessarily in the same place they were before. The amount of visceral fat can be expanded in adults via proliferation of those fat cells, but permanently decreasing fat cell number by weight loss has not been documented. It is an unfair one-way street: you can gain more fat cells, but as far as we know, you can never lose those fat cells no matter how diligently you diet or exercise.

Diligent and stoic adherence to a restrictive diet and a vigorous exercise regimen can successfully shrink, or even empty, existing fat cells. There is no evidence that empty fat cells automatically undergo cellular suicide (apoptosis). It would not make good evolutionary sense for this to happen anyway because healthy fat cells would be required for the organism to survive periods of famine or fasting. Moreover, it is likely that shrunken fat cells would "crave to be filled" because expression of the satiety hormone, leptin, closely parallels fat mass, and small fat cells secrete the least leptin. This means that the more fat cells you have, the harder it will be to succeed at long-term weight loss; therefore, the sooner you can reduce your contact with obesogens, the better.

A staggering 83 to 87 percent of those who work extremely hard to achieve significant weight loss regain the weight within a few years.[38,39] Why would they work so hard to achieve major weight loss, only to forget what they had learned and revert to their original, fatter self? This does not make much sense, except viewed through the lens of altered metabolic set points. A very recent study of people who lost massive amounts of weight during a season of *The Biggest Loser* and then regained it showed that these contestants had metabolic set points that were permanently programmed for the worst.[57] So much so that in order for them to maintain their weight loss, they would have to dramatically and permanently restrict their caloric intake and exercise many hours a day.

In my lab, Raquel Chamorro-García tested the effects of changing diet composition on F4 mice, whose F0 ancestors had been exposed to TBT throughout pregnancy and lactating.[77] At nineteen weeks of age, both TBT and control mice had about the same percentage of body fat on a normal, low-fat diet (13.2 percent calories from fat). She then switched the mice to a slightly higher-fat diet (21.6 percent calories from fat), although this still qualifies as a low-fat diet. Remarkably, the TBT mice gained weight very fast on the new diet compared with controls, becoming obese within six weeks. Moreover, when she fasted the animals, those whose ancestors had been exposed to TBT lost less fat than did the control animals. In short, the TBT animals handled calories differently; they gained weight more quickly and then resisted losing fat even during fasting—every dieter's lament. We do not yet know precisely how, but understanding which genes we have altered and in what way they are changed is the top priority in my research at the moment. Our working hypothesis is that obesogen exposure causes large-scale changes in how DNA is arranged in the nucleus, which leads to altered DNA methylation resulting in leptin resistance and a predisposition to higher expression of obesity-related genes.[90] In other words, this changed DNA structure leads to certain biological "switches" being turned on or off, predisposing the TBT animals to store fat and to resist mobilizing it.

In 2015, a team of researchers at Yale University School of Medicine led by Matthew Rodeheffer published a study that further revealed how the body can respond to a high-fat diet. They showed that in contrast to what had been believed, a high-fat diet can induce the production of more visceral fat cells, even when the existing cells are not yet full.[101] Visceral fat cells are the belly fat cells that surround the liver and other abdominal or visceral organs, such as the kidneys, pancreas, heart, and intestines. As we discussed in chapter 2, this type of fat is the most devastating to our health. In their paper, Rodeheffer and colleagues explain that when the mice were fed a high-fat diet, their visceral fat mass increased and they formed new fat cells ***before*** existing fat cells were filled up. This finding was opposite to the common dogma that new fat cells are not formed until existing cells are filled. I am fairly certain that future studies will show that obesogens have the same effect as a purely diet-induced obesity—increasing the number of visceral fat cells. And as we will see in chapter 5, obesogens can also change the intestinal microbiome to further disrupt metabolism and balance of fat storage in the body.

THE CHALLENGES OF OBESOGEN RESEARCH: WHY CAN'T WE KNOW MORE AND REGULATE MORE?

As we scientists continue to amass data on the effects of obesogens in laboratory tests and in animal models, a good question remains: How are these compounds affecting humans? As I mentioned, getting high-quality, long-term data on the relationships between chemicals and obesity in humans is challenging. Not only is it expensive and time-consuming, but in many respects it is impossible to unequivocally demonstrate cause and effect because, as mentioned, it is unethical to conduct controlled exposures of humans to toxic chemicals as we do for drugs. The best evidence for the effects of chemicals on humans comes from accidental exposures, as noted earlier, and from experiments using animal models. As a result, regulators struggle to determine acceptable exposure levels for the chemicals, and the obesogen field remains on the sidelines of environmental policy and clinical practice.

The Toxic Substances Control Act of 1976 (TSCA) is old and out-of-date; it has not kept up with science, and the vast majority of chemicals that are approved for use under the TSCA were never tested for safety in any way, let alone for endocrine activity. Chemicals in use when the TSCA was passed were "grandfathered"—that is, they were assumed to be safe without testing to prove it. Its replacement, the Frank R. Lautenberg Chemical Safety for the 21st Century Act (S.697), is worse in many ways. Perhaps most egregiously, it removes the power of individual states to regulate chemicals, instead mandating a uniform, nationwide policy that will be approved by the EPA, should the EPA continue to exist in the new political climate. This is a major victory for chemical companies who lobbied hard for this provision, but a catastrophic blow for chemical safety. Why do I say this?

The EPA has not tried to outlaw a chemical under the TSCA since the 1980s. Since the TSCA was enacted forty years ago, the EPA has banned only five existing chemicals (of more than eighty-six thousand in common use) as posing unreasonable risk to human health. These are polychlorinated biphenyls (PCBs), chlorofluoroalkanes (CFCs), dioxin, asbestos, and hexavalent chromium (made famous by Erin Brockovich). However, even these efforts have not been successful, since the asbestos industry successfully challenged the asbestos ban. The EPA has a program intended to test chemicals for endocrine disruption, the Endocrine Disruptor Screening Program, which was mandated by the Food Quality Protection Act of 1996 and an amendment to the Safe Drinking Water Act passed in the same year. But the first orders for initial testing of chemicals for endocrine disrupting activity were not issued until 2009. Worse yet, many of the approved tests were developed decades ago and are relatively insensitive to EDC action. Although scientists in the EDC research community have identified nearly one thousand chemicals with endocrine disrupting activity, the chemical industry disputes these findings, and as of September 2017, the EPA has not banned or regulated a single chemical based on its endocrine disrupting activity. Interestingly, the FDA has banned the antibacterial agent triclosan because it is an EDC. This is a sorry state of affairs.

Unlike other nations, the United States operates under the policy that a substance is deemed safe until proven otherwise. That is, the burden is on the EPA to prove that a chemical is hazardous to humans before it can be banned. While it may make sense to allow a hundred guilty men to go free rather than send one innocent man to jail (the origin of the presumption of innocence concept), it makes no sense at all in regulating chemicals. This position stands in stark contrast to the policy in Europe, where the burden of proof is on the company to prove that a chemical is safe before it can be licensed. Therefore, it behooves us to operate under a totally different personal policy—one that I will be outlining in part 2.

But before I get to that personal policy, there is one more important piece of the obesogen story that must be addressed: the power of epigenetics. In the past few years, research has exploded on how behavioral and environmental factors change how our genes are expressed. Obesogens are probably not mutating our DNA, but they could be changing how the genes encoded by our DNA are being expressed. Let me explain.

CHAPTER 4

The Power of Epigenetics

What Obesity and Other Non-communicable Diseases Have in Common

I am frequently asked how much *X*, *Y*, or *Z* contributes to a condition such as obesity. In other words, how much do genetics, dietary preferences, or behavior contribute to someone being overweight or obese? Sugar, for example, has gotten a lot of bad press recently. Some have argued that sugar is the main culprit in our obesity epidemic and that we can trace the rise in sugar consumption to the rise in that number on the scale, as well as in other health challenges.

The truth is we do not know the exact "ingredients" to obesity, especially on an individual level. We may never know precisely what drives person A into obesity while person B remains at a normal weight. The question is made more difficult by the incredible variation among individuals, which is becoming increasingly apparent in studies of diet and metabolism.[19] Person A may consume twice the amount of sugar as person B but never have an issue with weight and its related health problems. We would need to follow tens of thousands of people from preconception throughout adulthood while measuring many parameters such as what they eat, how much they exercise, where they live, what chemicals they were exposed to in utero and during early life, and so on in order to answer that question. We almost had a "National Children's Study" in the United States that would have addressed many of these issues, but unfortunately, it was abruptly canceled in 2014 amid allegations that it was poorly planned and had flaws in design.[102] National Institutes of Health (NIH) director Francis Collins suggested that the study would emerge from the ashes in a new form, but this has not happened yet. Fortunately, there are equivalent studies under way in Japan and elsewhere that will help us to understand the relative contributions of different factors to obesity and other health concerns.

Irrespective of how many factors contribute to obesity, one thing ***is*** certain: your DNA is not your destiny. That is, your health does not depend simply on what genes you were born with. The new science of epigenetics reveals how your environment (broadly defined) and the choices you make can change how your genes are expressed. In some cases, these changes can be passed on to your children, grandchildren, and beyond. We often hear people say, "I take after my mother" or "I'm built like my father." Obviously this must be true to some degree, because you inherit one copy of all your genes from your mother and another copy from your father (except in boys, who can inherit their X chromosomes only from mom and their Y chromosomes only from dad). But more correctly stated: Your life has been directly influenced not just by the genes you inherited, but also by what you were exposed to while your mom was pregnant and during your formative years, as well as the experiences of your parents and grandparents—what they ate, how they lived, and what they experienced. In this view, your health was at least partly influenced before you were born and maybe even before your parents were born.

The debate between nature (your DNA) and nurture (your environment) has persisted throughout modern history, particularly among psychologists. However, it is clearly a contrived debate—today, it is obvious that both genes and environment are important, as are interactions between genes and environment. The toxicology community has jumped into this type of contrived debate. In 2017, I was invited to an industry conference to debate the topic "Which is more fattening, the pizza or the pizza box?" Obviously both the pizza and the pizza box are fattening, but for different reasons. One hopes that the organizers were aware of how silly and irrelevant such a debate would have been, but you never know.[iv] Both nature and nurture have long been considered for their influence on the incidence rates of diseases, from obesity to cancer. About 18 percent of diseases have been associated with specific, genetic causes (for example, sickle cell anemia and hemophilia). The rest are multifactorial (require the action of multiple defective genes) or unexplained, despite the enormous amount of effort that has been expended into linking

---

[iv] The toxicology community is fond of such contrived debates. For example, I was invited to the Society of Toxicology Annual Meeting in 2015, where one night they staged a debate about whether low-dose, non-monotonic dose responses existed or not. Of course the audience voted that they did not exist because industry and government toxicologists, who made up the bulk of the audience, believe that all dose responses should be linear and that there is always a dose (usually a high dose) below which no adverse effects can be detected, as discussed elsewhere in the book. While one might not see the harm in a spirited debate, facts are not debatable, only theories and opinions are. Senator Daniel Patrick Moynihan put it best when he said, "Everyone is entitled to his own opinion, but not to his own facts."

genes to diseases around the world. One explanation would be that we simply have not tried hard enough to find the disease-causing genes, despite having sequenced the genomes of humans and all the major groups of mammals and after hundreds, if not thousands, of so-called genome-wide association studies (GWAS) that attempt to link DNA sequences with disease incidence. A better explanation would be that there are epigenetic factors that have not yet been considered adequately. What do I mean by that? Before we go there, it will help to begin with a short primer on genetics so that we are all on the same page.

THE DISCOVERY OF DNA HERALDS A NEW ERA

The twentieth century was no doubt the century of the genome with the discovery of DNA (deoxyribonucleic acid, the genetic material that underlies most life on earth). But the story of the genome goes much further back in history, and a number of books have been written on the topic.[v] The Austrian Augustinian monk Gregor Johann Mendel was a passionate gardener in the nineteenth century who gained posthumous fame as the father of genetics because his careful experiments predicted the existence of discrete units of heredity that gave rise to particular traits. Breeding peas in the garden at St. Thomas's Abbey, Mendel noted the effects of crossing different strains of the common pea plant. He demonstrated the transmission of characteristics

---

[v] Perhaps the most accurate story about the science of the genetic material and all of the players involved was related by Horace Freeland Judson in his classic book, *The Eighth Day of Creation*,[103] but it is also useful to read James D. Watson's *The Double Helix*[104] and Francis Crick's *What Mad Pursuit*[105] for a full appreciation of all the challenges and successes along the path to discovering the structure of DNA.

in a predictable way by inherited "factors" that would later be called genes. He showed that characteristics get passed on from one generation to the next, and that for each characteristic in an individual, each parent contributed to that characteristic. He further showed that genes could have variations, resulting in dominant and recessive traits. Mendel's work was not widely appreciated during his lifetime but was rediscovered in the early 1900s and refined into what we now know as Mendelian genetics by the work of geneticists such as Thomas Hunt Morgan.

While it became well accepted that there were units of heredity termed genes, it was not known what type of molecule (DNA, RNA, or protein) served as the "genetic material." Oswald Avery, working at what was then the Rockefeller Institute (now Rockefeller University), showed conclusively in the 1940s that DNA, rather than protein, was the genetic material. It was known that natural DNA contained phosphate, deoxyribose sugars, and nitrogen-containing bases (purines and pyrimidines), but the structure of DNA and how it could be replicated accurately were not known. Perhaps the next big breakthrough was made by the Austrian chemist Erwin Chargaff, who made the next step by determining how the "ingredients" to DNA matched up chemically. Working at Columbia University, Chargaff showed that in any naturally occurring DNA, the amounts of purine bases (cytosine and guanine; C and G) and pyrimidine bases (adenosine and thymidine; A and T) were roughly equal. That is, the amount of A was equal to the amount of T, and the amount of C was equal to the amount of G, which became known as Chargaff's first rule (A=T, G=C). Why this was so went unexplained until Francis Crick and James Watson, working at Cambridge University, together with Maurice Wilkins at King's College London, deduced the structure of DNA.[106]

Watson and Crick, with an assist from Wilkins, are most commonly identified as the ones to discover the structure of DNA, but their work built on the work of many others, including Rosalind Franklin, also from King's College, whose X-ray photographs of DNA structure led Watson and Crick to their "aha moment." What Crick, Watson, and Wilkins brought to light in the mid-1950s, which earned them the Nobel Prize in Physiology or Medicine in 1962, was that DNA was organized as an antiparallel double helix. This double helix had a phosphate backbone, held together by bonds between the 5' and 3' carbon atoms of successive deoxyribose sugars. The purine and pyrimidine bases extended out from this double helix, and the two strands were latched together by hydrogen bonds between opposing A and T or G and C bases.

Imagine that you had a flexible wooden ladder, held the ends, and twisted it and you will have something like the basic structure of DNA (not exactly, but close enough). In this model, the paired bases serve as the rungs of the ladder. This structure immediately suggested a mechanism for duplicating the DNA molecule—that each strand could serve as a template for a new one. The nature of how this happens was deduced by Matt Meselson and Franklin Stahl at the California Institute of Technology, who showed in 1958 that each DNA double helix unwinds and serves as a template for a new, complementary strand.[107] This way DNA can be precisely copied without changing its structure, with the exception of occasional errors or mutations. The DNA code is read as the sequence of A, C, G, and T nucleotides that constitute each strand of the DNA. The nucleotide sequence is a key structural element for the genes that individually, or in combination, determine everything from your hair color to your predisposition for certain diseases.

Watson and Crick's *Nature* paper transformed the life sciences and ushered in the era of molecular genetics—suddenly, learning the underlying DNA sequence seemed very important. It was widely assumed that once we knew the sequence of the entire human genome, it would be possible to understand virtually everything about how the human body worked, what genes were responsible for our individual traits, disease susceptibilities, and so on. However, it took some time for the technology required to determine individual DNA sequences to be developed. Walter Gilbert at Harvard University and Fred Sanger from Cambridge University shared the 1980 Nobel Prize in Chemistry for developing the first robust methods for determining DNA sequences. This opened the door for sequencing the genomes of viruses, bacteria, and so on.

The pace of genome sequencing has been breathtaking: from the first genomic sequence in 1978 of a small virus called ΦX174 (5,386 bases encoding eleven genes), which infects bacteria (commonly called a bacteriophage), to the "complete" sequence of the human genome, roughly three billion base pairs, in 2003. It took a mere twenty-five years from the first sequence to the human genome and only fifty years from Watson and Crick's *Nature* paper describing the structure of DNA.[106] There continues to be a concerted effort to determine which DNA sequences put one at a greater or lesser risk for diseases of various sorts, but there is still much to be learned about how genes function and how genes interact with the environment. The costs to sequence DNA have come down significantly in the past decade with the advent of better technologies, falling from almost $10 million in 2008 to close to $1,000 today (the first human genome took about $2.7 billion to complete). This was largely driven by the "$1000 genome" project funded by the NIH.

DNA AND DISEASE

We have grown accustomed to thinking that our individual DNA sequences almost completely control our health and wellness—even among members of the medical community. But this view is an oversimplification, if not largely mistaken. DNA sequence is just one part of the puzzle. DNA says more about our risk than our fate. DNA sequence controls probabilities for the most part, not necessarily destinies. Of course, any one of us can have a mutation or gene that encodes an absolute outcome, such as hemophilia, cystic fibrosis, or muscular dystrophy. But those types of conditions are rare because the majority are "recessive" mutations. That is, development of the disease requires two mutant copies of the gene, one from each parent. Dominantly inherited diseases—those that are caused by a single mutant copy of a gene—are much less common, with the exception of genes that are encoded on the X chromosome in men (such as red-green color blindness), because men have only a single X chromosome.

One way to get an idea of your potential risk for any genetic disease is to subscribe to any of the large number of personal genetic testing databases and see how many traits you are at risk for. I predict that most readers will find very little, if anything. I did "23andMe" before the FDA arrogantly prevented them from telling you much about your personal disease profile in 2013. I am happy to report that, for what it's worth, I have no major susceptibilities for any disease or condition (ranging from about half of normal to twice normal for a variety of conditions, but most in the plus/minus 50 percent range).

Your risk of this or that trait might be about 1.4 times higher or lower than "normal," but it is very unlikely that you will discover many DNA sequence variations strongly linked with any disease. Having said that, I have a friend who discovered from her 23andMe profile that she had a mutation in a gene that is linked to progressive hearing loss, and sure enough, she had noticed that it was becoming progressively harder for her to hear seminar speakers for the past few years. So the take-home message is that these types of testing services have some value, but what they tell you will most frequently be that your susceptibility for some condition will be slightly higher or lower than normal. This might call for some modest lifestyle modifications, but rarely more than that.

Another thing that it is virtually impossible to explain in the context of alterations to our DNA sequence is the staggering rise in the frequency of non-communicable diseases (NCDs) in the past thirty years or so.[108] By non-communicable, I mean diseases that are not caused by infectious agents such as viruses (for instance, colds and influenza), bacteria (cholera, tuberculosis, pneumonia), fungi (cryptococcal meningitis, "Valley fever"), or protozoans (brain-eating amoebas). According to the World Health Organization, a whopping 70 percent of global deaths in 2015 were due to NCDs.[109] The rise in NCDs is especially puzzling considering the sheer number of transformative medical breakthroughs over the last sixty or so years. The discovery of antibiotics, widespread recognition of the value of environmental hygiene, and increased use of vaccines against devastating diseases such as polio provided a great boost in life span. Other developments aided this life extension, including improvements in diagnostics and medical care, a decline in smoking, and increased accessibility to medical care and pharmaceuticals. Sadly, we have taken a sudden turn for the worse. For the first time in more than two decades, life expectancy for Americans declined in 2015. The report released by the National Center for Health Statistics in 2016 cited rising deaths from heart disease and stroke, diabetes, accidents, drug overdoses, and other conditions.[110] A very troubling study showed that life expectancy at five years of age in mid-Victorian England (1850–1870) was as good as or better than that today if one eliminated infectious diseases from consideration as a cause of death.[111] On top of that, the incidence of degenerative diseases such as cancer, cardiovascular disease, diabetes, and the like in mid-Victorians was 10 percent of what we experience today. Despite all our technological advances, the horizon is looking fairly bleak at the moment and we need to understand why this is.

Non-communicable diseases are now the number one cause of death in the world. The World Health Organization calls NCDs "a slow-motion catastrophe." Some astonishing numbers:

- Leukemia and brain cancers: over 20 percent increase since 1975.
- Asthma: doubled between 1980 and 1995, has remained elevated.
- Autism: increased 1,000 percent in past three decades.
- Infertility: 40 percent more women had difficulty conceiving and maintaining a pregnancy in 2002 than in 1982 (doubled in women aged eighteen to twenty-five years).
- Autoimmune disorders: according to a new study, the prevalence and incidence of autoimmune diseases, such as lupus, celiac disease, and type 1 diabetes, are on the rise, and researchers at the Centers for Disease Control and Prevention (CDC) are unsure why.

- We already mentioned type 2 diabetes, but also: between 2001 and 2009, the incidence of type 1 diabetes increased by 23 percent, according to the American Diabetes Association.

A century may seem like a long time compared with our usual life span, but it is just a moment on an evolutionary scale. Thirty years is an evolutionary instant. There is no possibility that human genetics has changed throughout the world rapidly enough to account for changes that happened over a thousand-year time frame, let alone a thirty-year time frame. This is unequivocal evidence that something other than alterations in the sequence of our DNA is leading to the worldwide increase in NCDs. One common explanation we hear is that our modern lifestyles are to blame. If you travel as much as I do, you will clearly see that this does not work either, because lifestyles in various countries are very different. Another striking way to think about this new reality is to consider that infant obesity almost doubled in a mere twenty years. We can't really blame six-month-old infants for "unhealthy lifestyles." Nor can we use the standard explanations of unhealthy lifestyle for the epidemic of obesity in domestic cats and dogs, urban feral rats, and five other species of animals that David Allison and his colleagues observed.[14] (See chapter 1.) We do not know what is triggering the rapid rise of type 2 diabetes, childhood asthma, autism, attention deficit hyperactivity disorder (ADHD), and infertility, but it is certain they must be due to environmental, dietary, and behavioral factors. Clearly, disease is more than just genes. The environment and how our genomes and "epigenomes" interact with the environment have prominent roles to play.

THE POWER OF EPIGENETIC EFFECTS

Do you remember when you first heard about evolution in biology class? Perhaps you even remember hearing about Charles Darwin's theory of "natural selection" compared with Jean-Baptiste Lamarck's theory that characteristics one acquires in life could be transmitted to one's offspring. Lamarck was the French naturalist who, you'll recall, proposed in 1802 that the environment can directly alter phenotype in a heritable manner—a mechanism for evolution in which species pass traits acquired during their lifetimes to their offspring. The example I gave earlier and that you most likely read about in school describes how antelopes could have evolved into giraffes by stretching their necks to reach the higher leaves on a tree. The "stretchiest" giraffes would pass slightly longer necks to their offspring. Every generation would lead to a slightly longer neck until the antelope was now a giraffe. Of course, if this were true, one wonders why the increases in neck size stopped at giraffe length instead of continuing to increase as the lower leaves were eaten.

This theory of evolution differed from Charles Darwin's later thesis that organisms cannot alter their genetic material on demand as needed. Whereas Lamarck's premise says adaptations appear as needed in response to the environment and the acquired traits are then passed on to offspring, Darwin's theory of evolution by natural selection holds that evolutionary changes in organisms result from differential procreation or survival in response to a changing environment. That is, natural selection acted on preexisting variations in the population. The organisms that were best equipped to prosper in current environments would produce more and fitter offspring, eventually dominating the population. Those organisms harboring variations that made them less fit eventually became underrepresented in the population. In this view, antelopes with the longest necks would have better access to food, enabling them to reproduce more, and eventually replace the short-necked ones.

Most biologists dismiss the possibility of Lamarckian inheritance since it doesn't immediately make sense and there was no ready mechanism to explain it.[vi] While Darwin's theory of natural selection[vii] has long dominated evolutionary theory, there is now room for a version of Lamarck's ideas. As is often the case in science, very smart people are rarely completely wrong. It is hard to believe that antelopes grew their necks as Lamarck proposed, but what about the effects of a changing environment on living populations? Wouldn't it make evolutionary sense if organ-

---

[vi] The Soviet biologist Trofim Lysenko fomented his own version of Lamarckism and used his influence with Josef Stalin to impose his views on Soviet science. Lysenko rejected the concept of genes, of Mendelian inheritance and natural selection, and his influence with Stalin led to the jailing, firing, or execution of scientists opposed to his teachings. Lysenko's version of Lamarckism essentially led to the destruction of genetics in the Soviet Union. It was not until 1964 that Lysenko's influence was finally removed. As you can easily imagine, Lamarck's theory was conflated with Lysenko's and, rather than being treated as an alternative theory for consideration and scientific debate, was emphatically dismissed along with Lysenko's twisted version. Mike Skinner has written a very thoughtful article describing a unified theory that accommodates both genetic and epigenetic inheritance.[92]

[vii] Niles Eldredge and Stephen Jay Gould introduced the concept of punctuated equilibria to make the Darwinian idea of natural selection more consistent with the fragmentary nature of the fossil record. In their view, populations had lots of variability but were stable over time (that is, in stasis or equilibrium) until some large environmental disturbance—the punctuation—led to the seemingly rapid changes observed in the fossil record.[112] One example of such a large-scale change would be the asteroid impact that coincided with the mass extinction of the dinosaurs about sixty-six million years ago.

isms also had a rapid way to respond to environmental changes? Lamarck got the details completely wrong but was probably correct in a broad sense—epigenetic inheritance ***is*** essentially the inheritance of acquired characteristics.

It is important to note that while evolution by natural selection is well supported by the fossil record and experimental data, Darwin also got some details wrong. For example, Darwin believed in a "blending" model of inheritance that was fashionable in the mid-1800s. Basically, blending inheritance holds that variation in individuals was random but bounded by the traits of the parents. The "blood," or hereditary traits, of both parents came together in the offspring, just as two colors of paint blend when mixed. While there are some instances where this might be true, it completely fails for continuously graded traits such as height. For example, a tall father and short mother would have children that were somewhere in between the two heights. If blending inheritance was the underlying mechanism, then offspring in every generation would always be less extreme than their parents for every trait and would approach an average over time. This is completely inconsistent with what we see in the real world and with the idea of gradual evolution by natural selection. Blending was an idea that Darwin wrestled with in his writings.

> Epigenetics is the study of changes in gene function that do not entail changes in the underlying sequence of DNA. These changes may be inherited when cells divide, or even across generations. Our environment and experiences can lead to modifications of DNA that change not the sequence, but how it is expressed. These chemical "marks" in the epigenome alter the function of genes, turning them on or off, or increase or decrease their expression.

The term "epigenetics" (literally, "on top of genetics") was first coined by the British developmental biologist Conrad Hal (C. H.) Waddington in 1942. Waddington had proposed a theory of "genetic assimilation" to explain some intriguing experimental results he had seen. Genetic assimilation referred to a trait (phenotype) that was originally produced in response to an environmental condition but that later became fixed in the genome. Waddington believed that a cell during development was like a ball rolling down a hill that was filled with gullies. Near the top of the hill, the ball rolled into one or another gully, perhaps as a result of hitting a rock or being displaced by a gust of wind. Once the ball was in a gully, it was difficult to get back out—not so different from embryonic cells, which start out with the ability to form all of the types of cells in the body (what we call pluripotent) but are later restricted in their ability to transition from one cell type to another (for instance, from muscle to brain). The influences acting on the cell were "epigenetic" since they were on top of the existing genetic information. Waddington found that he could induce extreme phenotypes in fruit flies (an extra set of wings) by treating them with ether and that he could breed these flies and increase the frequency of these effects until they could be seen in the absence of ether treatment. Although Waddington showed that these changes became more stable over successive generations as a result of selection, he was strongly criticized by colleagues such as the evolutionary biologist Ernst Mayr, who believed that Waddington was invoking Lamarckian inheritance. Waddington was prescient and his ideas have found support in the work of Mike Skinner, in our work, and in that of other laboratories.

Epigenetics simply refers to changes in gene expression ***without changes to the underlying DNA sequence***. Unlike standard genetics, which studies changes in the sequence of the DNA letters (A, T, C, and G) that make up our genes, epigenetics examines changes that do not alter the sequence of the DNA code. Rather, epigenetics changes when and how genes are expressed. Epigenetic "tags" play an important role in whether chromatin—the material of which our chromosomes are composed—is accessible to the complex of proteins that controls gene expression (known as "the transcriptional machinery"). If the chromatin is not accessible, nearby genes will not be expressed, even if all of the other necessary factors are available. These epigenetic tags can be added, removed, or changed in response to environmental factors. In turn, the presence or absence of these tags can play a key role in whether a gene is expressed or not. Think about it for a moment. A gene can be rendered nonfunctional by a mutation (change in the DNA sequence) that causes a truncated or defective protein to be produced. An epimutation (change in the epigenetic tags) that prevents a gene from being expressed causes exactly the same effect—absence of a functional protein—but by an entirely different mechanism. Or an epimutation can lead to a gene being expressed at a time and place when it would normally not be expressed. Either of these can alter how an organism functions and responds to its environment.

Epigenetic forces help explain why identical twins can grow up to look and behave somewhat differently from each other and why each possesses a different assortment of risk factors, despite harboring precisely the same DNA sequence. Their DNA may be identical, but the sum of epigenetic changes that they acquire in response to their environment throughout life alters how their identical DNA is expressed. Want an example? While more than 90 percent of identical twins are very close to the same height, less than half of identical twins will share traits such

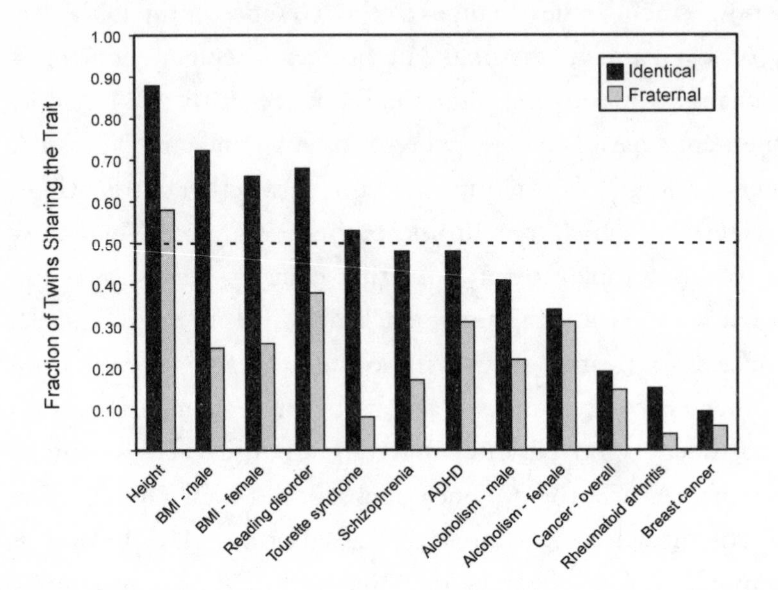

as alcoholism, diabetes, breast cancer, and rheumatoid arthritis. This provides conclusive evidence that epigenetic effects can be profound.

The field of epigenetics has only recently gained traction thanks to advances in DNA-sequencing technology that allow us to detect these "epimutations." Although we have known for quite a while now that the environment can affect gene expression, a new and provocative finding is that these epigenetic effects induced by chemical exposures can be passed on to future generations. Mike Skinner of Washington State University was the first to show this,[91] but my lab[90] and other labs[113] have also shown that environmental effects on physiology (in our case, obesity) can be passed on to future generations.

What all this means is that some of your genes, right now, could be behaving in ways inherited from your parents and even grandparents. What we eat, how much we exercise, where we live, whom we interact with, how much sleep we get, and even the aging process all can eventually cause changes in the epigenome around important genes that modulate whether and to what extent these genes are expressed over time. We refer to the changes in the epigenome as epigenetic "tags" or epimutations and the type of inheritance this produces as epigenetic.

I should point out that compared with genetics, the field of epigenetics is still in its infancy and continues to suffer from the anti-Lamarckian bias in genetics. The heritability of epigenetic changes, particularly those induced by environmental factors, remains controversial because no one has convincingly proved the underlying mechanism, although our work suggests that at least some effects of EDCs result in long-range changes in the structure of DNA that affects the expression of many genes, and Mike Skinner's work has identified specific changes in DNA methylation that may be heritable. Epigenetic inheritance is well documented in plants and worms, but less so in humans.

One prominent epigenetic mechanism is called DNA methylation, usually on cytosine nucleotides. A methyl group is a carbon atom surrounded by three hydrogens that can be attached to the 5 position of the cytosine ring, producing what we call 5-methyl-cytosine. The presence or absence of blocks of 5-methyl-cytosines affects the structure of DNA. Methylation in the parts of a gene important for its expression (the regulatory regions such as promoters and enhancers) can strongly influence whether the gene is expressed or not. Usually, more methylation in a gene promoter inhibits its expression, and vice versa. For example, both mutation of the *INK4A* tumor suppressor gene and hypermethylation of the *INK4A* promoter can cause malignant melanoma because both lead to loss of the INK4A protein.[114] While the concept of DNA methylation and the regulation of gene expression is well established, there are many who believe that DNA methylation cannot be heritable, at least not in mammals, because DNA methylation is mostly erased during the development of sperm and egg (so-called germ cell reprogramming). This is an exciting and controversial area of study at the moment.

Another important mechanism involves changes to proteins that help "package" DNA, such as histones, around which DNA is wrapped. Histone proteins can be modified in various places by the addition of methyl groups, acetyl groups (two-carbon groups), and phosphates. The particular combination of histone modifications is called "the histone code." The histone code can be fairly complex and provide precise fine-tuning of gene expression. Some histone methylation can be inherited, at least in mice, and this is another prime candidate to transmit heritable epigenetic changes.

A third type of epigenetic change involves the expression of what we call "non-coding RNAs" (or ncRNAs). These RNA molecules come in both small and large flavors that are not translated into proteins. Without digging deeply into the biochemical details of their function, these ncRNAs can alter gene expression and have been shown to have a critical role in normal development and biological processes. For example, the non-coding RNA *MIR31HG* interferes with the expression of the INK4A protein mentioned above; higher *MIR31HG* expression leads to lower INK4A expression in melanoma patients.[115] Whether and how changes in expression of ncRNAs for multiple generations remains to be determined.

The environmental factors that can influence gene expression (and presumably evolution and physiology) are wide and varied, from ecological parameters such as temperature and light to stress and nutritional details such as caloric restriction or high-fat diets. A multitude of environmental chemicals from phytochemicals—biologically active compounds found in plants—to synthetic toxicants can also influence observed characteristics and health. Future research will further help us understand all the steps involved from environmental triggers to inherited changes in chromatin structure and DNA expression. We also know that environmental factors can influence epigenetic tags in children and developing fetuses in utero. This is where my work has homed in on the impact that endocrine disrupting chemicals can have on a growing organism whose cellular programming could be altered permanently by epigenetic forces.

DEVELOPMENTAL ORIGINS OF DISEASE

London-born David Barker was among the pioneers to link chronic disease in adulthood to growth patterns in early life. In 1979, Barker became professor of clinical epidemiology at the Medical Research Council Environmental Epidemiology Unit, now the MRC Lifecourse Epidemiology Unit at the University of Southampton. As he was studying birth and death records in the United Kingdom, he noted a striking relationship between low birth weight and a risk for dying from coronary heart disease as an adult. Barker developed a hypothesis, eponymously named the Barker Hypothesis (aka the fetal origins hypothesis), that nutrition and growth in early life are important factors in determining whether a child will grow up to be more or less vulnerable to metabolic and cardiovascular disorders. Barker proposed that a baby born to a mother who was malnourished during pregnancy would be more susceptible later in life to chronic diseases such as diabetes, high blood pressure, heart disease, and obesity, because the fetus had adapted itself to a nutritionally poor environment. His observations culminated in a *Lancet* paper published in 1989 in which he reported that among 5,654 men from Hertfordshire, UK, those with the lowest weights at birth and at one year of age had the highest death rates from heart disease.[116]

Barker was not the first researcher to record the relationship between early life conditions and later disease. The Norwegian doctor Anders Forsdahl was the one who initially formulated the hypothesis in 1977, when he noted that a mother's living conditions during pregnancy and the first years of a child's life have an important impact on that child's risk of chronic disorders later in life, especially cardiovascular diseases.[117] As a community physician, Forsdahl had firsthand experience of poverty and a high incidence of cardiovascular disease. Barker is often credited with establishing the theory of fetal programming, but he was merely popularizing a hypothesis formulated by Forsdahl (although Barker does cite Forsdahl in his *Lancet* publication). In one of Barker's last public speeches, he stated: "The next generation does not have to suffer from heart disease or osteoporosis. These diseases are not mandated by the human genome. They barely existed 100 years ago. They are unnecessary diseases. We could prevent them had we the will to do so."

Barker had his critics and opponents, but his work ultimately inspired others to study the fetal origins of adult illnesses and disorders. Peter Gluckman of the University of Auckland and Mark Hanson of the University of Southampton showed how important the developmental period is in the future health of an individual. They went beyond the fetal origins model and pioneered a new area of medical research called the Developmental

Origins of Health and Disease, or the DOHaD paradigm.[118] They wrote a 2006 book, *Mismatch*, in which they explain why we suffer from "lifestyle" diseases now.[119]

Just what are the "lifestyle" diseases of developmental origin? That is, what are the diseases that can be affected by exposures during the fetal and developmental period? All the chronic ones I have already mentioned: cardiovascular and pulmonary maladies, including asthma; neurological ailments, including ADHD, autism, and neurodegenerative conditions such as dementia; immune/autoimmune conditions; endocrine diseases; disorders of reproduction and fertility; cancer; and, yes, metabolic disorders, including obesity and diabetes. And here is a critical fact: ***Developmental programming is almost entirely epigenetic because the DNA sequence is not changed.*** Put simply, epigenetic changes are the mechanism through which developmental origins of disease happen. Prenatal and early life experience programs the body throughout the rest of life.

ROBUST BUT FRAGILE AT THE SAME TIME

Development is a unique process. On the one hand, despite how complex living organisms are, with many opportunities for things to go wrong, development largely works well because there are built-in redundancies and backup plans for potential errors that could otherwise lead to serious malformations, defects, or problems. The famous German embryologist Hans Spemann (who won the Nobel Prize in Physiology or Medicine in 1935) said that development has both a belt and suspenders (to hold its metaphoric pants up). Cells have elaborate mechanisms to repair DNA as well as mechanisms to kill themselves in an orderly way if too many mutations are detected—a process called apoptosis. Having said that, I should add that development is also exquisitely sensitive during certain stages, during which small changes can have major outcomes. These are called "critical windows" of sensitivity. One entertaining example is in frog embryology—if you tilt a frog embryo at a critical time before the fertilized egg divides into two cells, you will get an embryo with two perfect heads; do the same thing thirty minutes later and the embryos will be completely normal. Perturbations that occur during critical windows can lead to dramatic birth defects. For example, 80 percent of babies whose mothers were prescribed thalidomide to control nausea from morning sickness during pregnancy were born with severe limb defects if mom took the drug between twenty and thirty-six days after fertilization (that is, in the first month or so of pregnancy). However, if thalidomide was taken by mom outside of this critical window, the children did not have these birth defects.[120]

Developmental perturbations can also lead to more subtle alterations that cause adverse health outcomes and increased risk for disease that may not be observable until years later. For example, the nutrition of an expectant mother and her exposure to chemicals and tobacco products can have far-reaching and permanent effects on her offspring. Reduced fetal growth, which is often one of the results of exposure to nicotine in utero, is strongly associated with chronic conditions later in life, such as heart disease, diabetes, and obesity. There are many good reasons why physicians recommend certain guidelines to pregnant women: avoid drug X, take prenatal vitamins containing folate, don't smoke or drink alcohol, avoid exposure to Zika-infected mosquitoes, and so on. Insults to fetuses can have devastating and permanent repercussions.

A shortcoming of many epidemiological studies is their

inability to test cause and effect, as we can in animal studies. Occasionally, a tragic industrial accident, war, or other unexpected event offers the opportunity to study cause and effect in a human population (cohort). Among the first studies to provide early, undeniable clues to a cause-and-effect relationship between the prenatal experience and lifelong consequences was the Dutch Famine Birth Cohort Study, which found that the children of pregnant women exposed to famine were more vulnerable to diabetes, obesity, cardiovascular disease, kidney damage, and other health problems as adults. Known as the *Hongerwinter* ("Hunger winter") in Dutch, the Dutch Famine took place near the end of World War II during the winter of 1944–1945 in the Nazi-occupied part of the Netherlands, especially in Amsterdam and the densely populated western provinces. A Nazi blockade cut off food and fuel shipments from farm areas, leaving millions starving. Rations were as low as four hundred to eight hundred calories a day—less than a quarter of what an adult should consume. Those who survived, some 4.5 million people, relied on soup kitchens until the Allies arrived in May 1945 and liberation of the area alleviated the famine. The children of women who were pregnant during the famine were smaller, as expected, considering the severe caloric restriction. Surprisingly, however, when these children grew up and had children, ***those*** children were also shorter (but not lighter) than average and had increased fat and poorer health later in life.

The Dutch Famine Birth Cohort Study[121] was initiated by the Departments of Clinical Epidemiology and Biostatistics, Gynecology and Obstetrics, and Internal Medicine of the Academic Medical Centre in Amsterdam, in collaboration with the MRC Environmental Epidemiology Unit of the University of Southampton in the United Kingdom. The study has expanded to include other universities in the Netherlands (Leiden) and United States (Columbia University). Barker was in fact part of the study, which began publishing its results in 1998.

Data from this study showed that the famine experienced by the mothers caused epigenetic changes that were manifested later in life as a predisposition to disease and that some of these effects were passed to the next generation.[122] There are some data supporting the possibility that these effects are carried by DNA methylation, but this topic remains controversial at the moment. The DOHaD hypothesis suggests that the type and availability of nutrients during pregnancy and infancy have profound impacts on the individual's life. Another way to state this is to say that the effects of prenatal nutrition flow across generations. A baby develops from an egg in its mother that was created and nourished by the maternal grandmother during her pregnancy. This also means that a mother's exposures to hazardous chemicals while pregnant, and her baby's exposures while developing, can affect not only the future health of that child, but also the future health of that child's biological children. This is a sobering thought. It is one thing to suffer the consequences of our own choices, but quite another matter entirely to impose these effects on our descendants down the generations. I should also add that when it comes to epigenetic changes, both nutritional impacts (or deficiencies) and chemical exposures can be powerful forces.

An important distinction between the fetal origins hypothesis championed by David Barker and the DOHaD hypothesis formulated by Hanson and Gluckman is the realization that developmental programming does not stop at birth but continues throughout early life, probably at least until adolescence. One strong example comes from the famous Swedish Överkalix study.[123] This study is named after a small isolated municipality in northeast Sweden where nutrition of the local population was strongly dependent on the annual wheat harvest. What made this study possible were the detailed records available for harvest, food prices, births, and causes of death. The groups of Marcus Pembrey from University College London and Lars Olov Bygren from the Karolinska Institute found that drastic changes in nutrition during the prepubescent period (eight to twelve years old) affected longevity of grandchildren in the paternal lineage. Surprisingly and counterintuitively, when the paternal grandfather had an abundance of food (that is, was overfed) just prior to puberty, his grandsons had a fourfold increased risk for death from cardiovascular disease and type 2 diabetes. Conversely, when food availability was restricted during the father's prepubescent period (that is, they were undernourished), his grandsons were less likely than normal to die from cardiovascular disease and type 2 diabetes.

To summarize the somewhat complex science we have just discussed, the available results show that nutritional factors, chemical exposures, and stress during fetal development and at least until adolescence can lead to permanent effects on the health and well-being of individuals. In some cases, these effects can be passed on to future generations through mechanisms we are just beginning to understand. This is why Jerry Heindel talks about a good start (or a bad start) lasting a lifetime. When we think of malnutrition, the most frequent image that comes to mind is one of a skeletal child in the developing world with a protruding belly on the precipice of death. However, many children in so-called developed nations are suffering from a different but also harmful type of malnourishment: they are overfed with processed foods containing numerous EDCs from both agrochemical and industrial sources, while at the same time lacking essential micronutrients such as high-quality protein, healthy fats, and good carbohydrates including whole grains, fruits, and vegetables. They experience an excess of calories from products high in white flour and refined sugars and that lack real nutrients and vitamins. This is called "high-calorie malnutrition."

The fact that the human body, especially a young, developing one, is fragile but robust at the same time is critical, especially when we consider the impact of one particular class of chemicals known to reprogram biology and ultimately impact weight: obesogens. As with biological insults such as exposures to famine or too much food, we are now beginning to understand how obesogen exposure in utero and as newborns, toddlers, and even teenagers has the ability to alter gene expression, leading to a predisposition to obesity. But as you probably already know, obesogens alone are not to blame for our obesity epidemic. Other things can come into play as well, complicating matters. Let's go there next.

CHAPTER 5

It Gets Complicated

Other Fat-Inducing Factors

If only the story of obesity were exclusively composed of characters from the worlds of diet, exercise, and chemical exposures. Unfortunately, it is more complicated than that. As I have been hinting at all along, there are other factors that influence whether you are overweight or obese, independent of the energy balance equation. In this chapter, we will talk about seven important, interrelated factors, all of which interact with diet, exercise, and obesogens. They include the following:

- Your microbiome
- Consumption of sugar, including artificial sweeteners
- The quality of your sleep
- Your levels of stress
- Pharmaceutical drugs you take
- Exposure to certain viruses
- Inherited "fat" genes

Note that entire books could be written about each of these topics, so here I will summarize and highlight the essentials as they relate to your weight. Some of these topics may not seem directly related to obesogens, but they are. It is important for you to look at the impact of obesogens as a whole life problem—it is not a matter of just buying organic foods and seeking BPA-free bottles. By the same token, you cannot simply improve the quality of your sleep and exercise more to lose weight. To achieve permanent weight loss, you must evaluate your entire lifestyle and identify your personal vulnerabilities. Let's take each of these seven factors into consideration, starting with the friendly microbes that inhabit your body right now and have a say in everything about your weight equation.

MEET YOUR MICROBIOME

If you grew up in the twentieth century, then you probably came to understand microbes as small organisms invisible to the naked eye that can cause diseases in humans. Tuberculosis, bubonic plague, and whooping cough, for example, are all caused by microbes, specifically bacteria. You have no doubt heard about methicillin-resistant *Staphylococcus aureus* (MRSA) and necrotizing fasciitis (flesh-eating bacteria). Prior to the development of antibiotics, infectious diseases carried by bacteria were a major cause of death.

Only recently—in the twenty-first century—have we recognized that many microorganisms are, in fact, important for our health and vitality. The colonies of microbes, known collectively as the "microbiome," that inhabit our bodies—inside and out—and may outnumber our own cells by three- to ten-fold have stolen the spotlight lately. The human microbiome weighs about three pounds, the same as our liver or brain. The microbiome has become an important emerging new field of study since the pioneering work of Dr. Jeffrey Gordon at Washington University in St. Louis showed that the gut microbiome has profound impacts on our physiology and metabolism.[124,125] The microbiome comprises more than one hundred trillion single-celled microbes, mostly bacteria, that ride around with us on our skin and in our mouths, noses, intestines, and genitalia. The gut microbiome synthesizes vitamins, helps us digest food, bolsters our immune systems, and even boosts our brain function. Perhaps you will not be surprised to learn that a variety of lifestyle factors may play an important role in the composition of our microbiome and, hence, our health and well-being. Studies are currently under way to understand this complex relationship, but such factors may include the type of diet we eat (carbohydrate vs. fat, meat eaters vs. vegetarians, sugar vs. artificial sweeteners), how much we exercise, levels of stress, and what types of chemicals we are exposed to.[126]

Many microbiome projects are under way around the world that employ state-of-the-art genomic methodology to better understand how bacterial symbionts[viii] influence our physiology.[127] Large gaps in our knowledge about the interactions among diet, lifestyle, gut microbes, and health still exist, but by some measures, these collective projects may become more significant and game-changing in medicine than the human genome project. The gut has a separate nervous system, the so-called enteric nervous system, that integrates information about the state of the

[viii] A symbiont is an organism that lives together with another, a process called symbiosis. Symbiosis can be mutualistic, where both benefit, commensalistic, where one benefits but the other is not harmed, or parasitic, where one benefits and the other is harmed.

gastrointestinal tract to modulate how it functions. The enteric nervous system is strongly influenced by small molecules produced by gut bacteria, which means that they can have profound effects on what we eat and how we feel and behave.

The microbiome is a subject deserving of its own book(s), especially with respect to how it affects health and disease, from autoimmune disorders to cancers. For the purposes of our discussion, I will focus chiefly on how the microbiome impacts weight and obesity. Recent research has uncovered some stunning new clues to improve the weight loss endeavor, showing us that being overweight or obese may have more to do with the microbial profile in your gut than with your willpower to eat less and exercise more, or even your personal genetics. By changing the way we store fat, impacting how we balance levels of glucose in the blood, influencing the expression of genes that relate to metabolism, and affecting how we respond to hormones that make us feel hungry or full, our gut bacteria can have important influences on whether we lose or gain weight and how easy this is to accomplish.

Put another way, the actions of the microbiome further dismantle the "calories in versus calories out" equation when it comes to weight, because the microbiome plays a fundamental role in energy balance. The types of bacteria in our gut determine how much energy we extract from food as it passes through the gastrointestinal tract—that is, how much of the potential energy in food we consume is actually taken in and used. Animal studies show that if the microbiome has too many types of bacteria that efficiently harvest calories from food, more of the calories eaten will be absorbed, which can lead to increased body fat and, eventually, obesity. It is likely, although not yet proven, that the same occurs in humans.

In many cases, when a baby moves through the birth canal, it is "showered" by microbes that begin to colonize its gastrointestinal tract. Although there may be some exposure to microbes in the uterus via the placenta, the vast majority of the earliest colonizers come from the birth process. Those of us born via the much more sterile cesarean section tend to have different microbial profiles from those who are born naturally. It is not known yet whether the birth process itself compared with cesarean section alters the microbial populations so these differences are only an association.

Recent studies show that babies who are not exposed to a rich array of beneficial bacteria in early life will live with a higher risk for obesity and diabetes than their peers who develop healthier microbiomes.[128,129] The same holds true for babies who are exclusively formula fed, because they miss out on the healthy dose of protective antibodies and beneficial bacteria provided by breast milk and mom's skin. Early evidence of gut bacteria relating to obesity came from animal and human studies comparing intestinal bacteria in obese and lean subjects. The intestinal microbiomes of slender people appear to be much different from those of obese individuals. Peter Turnbaugh, Jeffrey Gordon, and colleagues convincingly showed that the microbiome was closely linked with obesity.[130] They started with genetically identical mice whose intestines completely lacked bacteria—so called germ-free (gnotobiotic) mice. Then they transplanted microbes from the intestines of obese or lean mice into the guts of these gnotobiotic mice. They found that the gnotobiotic mice adopted the obesity phenotype of the donor transplant—that is, the obese microbiome made the host mice fatter, whereas the lean microbiome kept the hosts lean, irrespective of diet.[130] Gordon and colleagues did similar experiments transplanting the microbiome from obese people or lean people into gnotobiotic mice and again found that the obese microbiome made the host mice fatter. The number and types of microbes in the obese vs. lean microbiomes were also different, with the obese microbiome much less diverse.

If this is not enough to convince you, they did a further experiment in which the microbiomes from obese vs. lean twin sisters were transplanted into germ-free mice.[131] As in the previous experiment, the mice receiving the lean microbiome stayed lean and those receiving the obese microbiome became about 15 percent fatter, despite eating the same diet. Remarkably, when the two types of mice were housed together in the same cages, the microbiome from the lean mice was transferred to the mice that had previously received the obese microbiome, preventing the latter from becoming fatter. This shows that the microbiome trumped both diet and genetics in controlling the fat content of these mice.

We also know that certain types of fats can affect the composition of the gut bacteria, which then influences whether an animal develops obesity-related inflammation. In 2015, a study published in *Cell Metabolism* showed that mice fed a high-fat diet composed mostly of lard for eleven weeks developed signs of metabolic disease, while mice that ate the same amount of calories from fat, but as fish oil, remained healthy.[132] When the researchers transplanted gut bacteria from fish-oil-fed mice to gnotobiotic mice and then subsequently fed the animals lard, the mice were protected from the usual unhealthy effects of the saturated fat. Previous studies had provided circumstantial evidence connecting the composition of the gut microbiome with food cravings produced by the hunger center of the brain.[133] This study, led by scientists at the University of Gothenburg in Sweden, went much further by revealing that feeding different types of fat resulted in very different gut microbiome profiles.[132]

Where the science really shines a light on the power of the microbiome is when we consider the effects of gastric bypass surgery on not just weight loss, but the almost overnight reversal of type 2 diabetes. Gastric bypass procedures involve making the stomach and small intestine smaller. Initially, it was thought that the rapid weight loss typically seen in patients who underwent the surgery was attributed to the individual eating less. But we now have evidence that a major portion of the weight loss is due to changes in the gut bacteria—changes that happen in response to the anatomical adjustments made by the surgery. It is unclear whether this results from dietary shifts that favor the growth of different bacteria or some other cause.[134,135]

Considering all of these studies together, it is clear that the quality and diversity of the gut bacteria factor mightily into metabolism and weight in animal studies and likely in humans as well, although the jury is still out. Explaining all the factors that contribute to the biochemistry and physiology of the microbiome and how these interact with physiology would take an entire book; plus, the story is far from complete. We will focus on two particular examples that illustrate the point: natural and artificial sweeteners—two related dietary villains that further demonstrate the importance of a healthy gut microbiome.

THE ROLE OF FRUCTOSE, SUGAR, AND ARTIFICIAL SWEETENERS

As mentioned earlier, fructose—a sugar naturally found in fruit and honey—has become one of the most common sources of calories in the U.S. (and increasingly the Western) diet. The most common sugar in nature, sucrose, is a disaccharide—one molecule of fructose chemically linked to one molecule of glucose.

But the majority of the fructose we consume is not in its natural form (that is, as part of sucrose) or source (whole fruit). The average U.S. citizen consumes 163 grams of refined sugars (652 calories) per day, and of this, roughly 76 grams (302 calories) are from a highly processed form of fructose, derived from high-fructose corn syrup (HFCS).[136] Professor Robert Lustig from the University of California–San Francisco has been sounding the alarm about sugars, particularly added, processed fructose, for many years now, as detailed in numerous scientific publications and his book *Fat Chance*.[137] Other experts, such as Dr. Michael Goran, director of the Childhood Obesity Research Center (CORC) and professor of preventive medicine at the University of Southern California, have argued that the amount of fructose we consume could be considerably higher given the murkiness, if not downright inaccuracies, in sugar labeling. Dr. Goran and his colleagues have discovered that the HFCS used in several popular beverages is delivering a level of fructose much higher than commonly thought. Although the manufacturers claim that sodas and beverages were made with HFCS 55 (55 percent of the sugar as fructose and 45 percent as glucose), Goran and his colleagues identified levels of free fructose as high as 65 percent in soda purchased around the Los Angeles area.[138] These results were published in the journal *Obesity* in 2011. Not surprisingly, the beverage industry responded by sponsoring studies showing that the HFCS 55 they tested contained close to 55 percent fructose.

Goran countered with a follow-up study in the journal *Nutrition* in 2014 where he commissioned three independent labs to measure fructose levels in a variety of beverages by different methods.[53] These studies confirmed the initial findings. Goran's team collected and analyzed the composition of thirty-four popular drinks and found that those made with HFCS contained 50 percent more fructose than glucose and that some product labels misrepresented the fructose content. For example, they found that while the label on Pepsi Throwback indicated it is made with real sugar (sucrose), in reality it contained more than 50 percent fructose. Sierra Mist, Gatorade, and Mexican Coca-Cola also were shown to have higher concentrations of fructose than reported by their labels. You can believe whom you choose, but I know Michael well and think that his studies are quite convincing.

Now, you might be wondering why the amount of fructose is such a big deal and thinking that 55 percent is not so different from 50 percent. Sugar is just sugar, after all, right? That is what the commercials for HFCS say. But HFCS is not the same as sucrose or the fructose that occurs in nature. In fact, many scientists have suggested that increased consumption of sugars in general, and fructose in particular, is contributing to the obesity epidemic; that fructose is a major factor in creating a so-called Westernized gut microbiome that lacks diversity and extracts more calories from food, essentially feeding your fat cells with a "thrifty metabolism."[139] We consume more HFCS per capita than any other nation,[136] and consumption has doubled over the last three decades as public health experts have urged us to cut fat consumption. Over the same time period, the incidence of type 2 diabetes has tripled. For these reasons alone, many experts make a connection between the increased diabetes (and obesity) and consumption of sodas, sports drinks, and energy drinks. Numerous studies now show that consuming fructose is associated with impaired glucose tolerance, insulin resistance, high triglycerides, and hypertension. As Goran wrote in a letter to former first lady Michelle Obama, "Although common table sugar (sucrose) is also comprised of glucose and fructose, nature balances them in equal proportions, and joins them by a bond for which the human body produces an enzyme (sucrase) to break down before absorption into the bloodstream. Therefore, the body absorbs fructose from sucrose more slowly than fructose from HFCS."[140] Natural fructose is found in fruits and vegetables together with dietary fiber. The absorption of this fructose into the bloodstream is blunted by this fiber. In contrast, high-fructose corn syrup contains free fructose, which disrupts liver metabolism and, along with excess glucose, elevates blood sugar levels and exhausts our pancreas.

Fructose is particularly troublesome in the body because it is primarily metabolized and stored in the liver, where it stimulates fat storage. Fructose also increases triglyceride levels in the blood and does not spike blood glucose or stimulate insulin production as does glucose. In turn, this means that your body does not produce leptin in response to fructose ingestion, so the body does not sense satiety, which can lead to increased food consumption. This same outcome—the lack of satiety upon consuming fructose—is also seen with artificial sweeteners. The human body cannot digest artificial sweeteners, which is why they have no calories. But the sensation of sweetness itself triggers some of the same biological responses as does sugar consumption, and the artificial sweeteners still must pass through our gastrointestinal tract.

For a long time we assumed that artificial sweeteners were, for the most part, inert ingredients in terms of affecting our physiology. Although it was first shown in 1988 that cyclamates, a class of artificial sweeteners, could modify how the microbiome behaved,[141] it was generally assumed that sugar substitutes such as saccharin, sucralose, and aspartame did not have a metabolic impact because they do not raise blood glucose levels. But it turns out that they can indeed wreak havoc (and cause the same metabolic disorders as real sugar) by triggering transient elevations in insulin levels (which increases fat storage) and alter the microbiome in ways that favor unhealthy biology. Studies are emerging to demonstrate that the gut bacteria of people who regularly consume artificial sweeteners is very different from those of people who do not. Consumption of artificial sweeteners has been linked with increased weight, higher fasting blood glucose, and elevated risk for developing type 2 diabetes. In 2013, French researchers published the results of a study that followed more than sixty-six thousand women since 1993: they found that the risk for developing type 2 diabetes was ***more than double*** for those who drank artificially sweetened drinks as compared with women who consumed sugar-sweetened beverages.[142] A recent meta-analysis[ix] of more than thirty studies, including more than four hundred thousand participants evaluated over ten years, revealed no benefit of artificial sweeteners for weight loss and found that long-term use of such sweeteners might be associated with increased BMI and risk for cardiometabolic conditions.[143] Clearly more research in this area is needed.

This discovery of a link between artificial sweeteners and the state of the gut microbiome stunned the scientific community when it landed in the journal *Nature* in 2014.[144] Eran Segal and Eran Elinav at the Weizmann Institute of Science in Israel, whom we met in chapter 1, led their team on a series of experiments to

---

[ix] A meta-analysis is a large-scale analysis that focuses on the results of previous studies. They are frequently performed on genome-wide association studies (GWASs), large-scale analyses that seek to link particular genetic regions with diseases, to derive more information from the increased statistical power of many studies.

answer one question: Do artificial sweeteners affect healthy gut bacteria? Segal, Elinav, and their colleagues started by adding the fake sugars—saccharin, sucralose, or aspartame—to the drinking water of different groups of mice. They gave other groups of mice the natural sugars glucose or sucrose in their water; the control group drank plain, unsweetened water. Eleven weeks later, mice that drank the artificial sweeteners exhibited signs of glucose intolerance compared with the controls. In other words, they could not control their blood glucose levels very well. This was an important finding, but where the rubber met the road in this experiment was when they tested the effects of artificial sweeteners on the microbiome.

To determine whether gut bacteria had anything to do with the link between drinking fake sugar and developing glucose intolerance, these researchers gave the mice high doses of antibiotics for four weeks to essentially eliminate all bacteria in their gut. Surprisingly, after a nearly complete elimination of their gut microbiomes, all of the groups were able to metabolize sugar equally well. As a final test of how fake sugars affect the microbiome, the researchers transplanted gut bacteria from mice that had consumed saccharin into gnotobiotic mice with no gut bacteria of their own. Within just six days, the now tainted mice had lost some of their ability to process sugar. Genetic analyses of the gut microbiome told the story: the saccharin microbiome was considerably different from the pretreatment microbiome. Some types of bacteria became more abundant, while others diminished, reminiscent of what Gordon and colleagues found for obese vs. lean microbiomes.[144] No doubt you are wondering whether this also occurs with your favorite artificial sweeteners in humans. These studies are under way but I will not be surprised if the results show that fake sugar has been faking us out for a long time.

Although we don't yet have any definitive studies to show the impact of environmental chemical obesogens on the microbiome, our preliminary studies, show that tributyltin, which we learned about in chapter 2, alters both the microbiomes and the metabolomes (the set of small chemicals produced as a result of ongoing metabolic processes in the body) of treated animals and their descendants. These studies in my lab and elsewhere are just beginning; future research will likely reveal which obesogens assault the health and functionality of the gut microbiome and how this in turn impacts healthy human physiology and metabolism. In the future, we should also be able to learn what we can do to preserve optimal intestinal ecology, from dietary strategies to the use of supplements such as prebiotics and probiotics. I would be careful about jumping onto this bandwagon right now, though. We simply do not know enough yet to reliably predict which bacteria in what balance promote optimum health. One thing we recognize can make us unhealthy and overweight is poor sleep. The research in this area—sleep medicine—is now extensive.

SOUND SLEEP PROMOTES A HEALTHY CIRCADIAN CLOCK

We all know people who claim to consistently get by on four or five hours of sleep. Indeed, after I presented our work at a symposium on nutrition, my colleague Dr. Francisco Ayala (one of the giants in genetics) noted that he had trained himself as a graduate student to function well on five hours of sleep a night so that he could outwork other researchers. Francisco is rail thin and wondered why, if sleep was so important, he was not fat. The answer: Who knows? Kudos to those who can get by on little sleep without this affecting their metabolism and overall risk for certain health challenges (not to mention feeling good and energetic). But for the great majority of us, sleep is more vital than we imagine, and most of us need a good seven to nine hours nightly. I average around seven a night myself, though sometimes I could use a little more.

Contrary to popular belief, sleep is not a state of inactivity. It is a critical phase during which the body replenishes itself in a variety of ways—ways that ultimately impact every system, from the brain to the heart, the immune system, and all the inner workings of our metabolism. Getting a good night's sleep (the details of which are coming up in part 2) is what keeps us sharp, creative, and able to process information in an instant as well as integrate new knowledge for future reference. Studies have repeatedly proven that sleep habits have the power to regulate the biological control systems that impact how hungry you feel, how much you eat, how efficiently you metabolize that food, how strong your immune system is, how insightful you can be, how well you can cope with stress, and how well you can remember things.[145-149]

Sleeping fewer than seven to eight hours in a twenty-four-hour period, or experiencing irregular sleep sessions during which there is inadequate time spent in deep, restorative sleep, has been shown to be associated with a spectrum of health challenges, from cardiovascular disease, high blood pressure, and diabetes to unintentional accidents, learning and memory problems, depression and other mood disorders, weight gain and obesity, and an increased risk of dementia, including Alzheimer's disease, cancer, and premature death. Dr. Matthew Walker, a professor of neuroscience and psychology at the University of California–Berkeley, used to say that sleep is the third pillar of good health, alongside diet and exercise. But given his research into the impact of sleep on the brain and nervous system, he now teaches that sleep is the single most effective thing we can do to reset our brains and bodies, as well as increase a healthy life span.[150,151] Several recent studies have revealed a possible mechanism for why this is so.

In 2012, Dr. Maiken Nedergaard and her colleagues at the University of Rochester showed that cerebrospinal fluid flow through the brain increased dramatically when mice were sleeping, but not when they were awake.[152] Cerebrospinal fluid is found in the brain and spinal cord, where it bathes and protects the central nervous system and eliminates waste products. They hypothesized that this flow might function like the lymphatic system in the body, draining tissues of cellular breakdown products and waste for eventual disposal. They named this the "glymphatic" system. In a follow-up study, they found that, in fact, this cerebrospinal fluid flow cleared toxins from the brain.[153] This has profound implications for our understanding of sleep—you must sleep in order for your brain to be cleared of waste products so that it can function properly.[154]

As with the parallel patterns of increased sugar consumption correlating with an increase in obesity, there has been a similar pattern in the increase in sleep deprivation and obesity. On average, Americans sleep about two fewer hours each night than they did a century ago. And only 15 to 30 percent of teenagers get the eight and a half hours a night recommended for them. It can be hard to appreciate the value of sleep, and for a long time we didn't really understand its purpose. Every animal sleeps, though, so there must be a logical, non-negotiable reason. In addition to its role in clearing the brain of toxins, sleep has a commanding role in our hormonal cycles, which in turn affect our basic physiology down to the speed and efficiency of our metabolism.

Every living organism has internal biological clocks called circadian rhythms. We have clocks in every cell in our body—a clock in our heart, muscles, liver, pancreas, and so on. These clocks regulate patterns of repeated activity associated with the environmental cycles of light and dark or day and night. These are rhythms that repeat roughly every twenty-four hours, and they include our sleep-wake cycle, the ebb and flow of hormones, and the rise and fall of body temperature that correlate with the twenty-four-hour solar day. They dominate all metabolism and physiology in most life forms, in fact. When your circadian rhythm is not synchronized properly with the twenty-four-hour solar day, you will not feel 100 percent. Most of you have experienced jet lag from airplane travel across multiple time zones or have suffered the effects of pulling all-nighters at work or in school. We know all too painfully what it means to have a disrupted circadian rhythm.

Put simply, your circadian rhythms depend a lot on your sleep habits. Healthy rhythms help regulate normal hormonal secretion patterns, from those associated with hunger cues to those that relate to stress and cellular recovery. Our chief appetite hormones, leptin and ghrelin, for example, orchestrate the stop and go of our eating patterns. The hunger hormone, ghrelin, tells us we need to eat; the satiety hormone, leptin, says we have had enough. We now have data to demonstrate that inadequate sleep creates an imbalance of both hormones, which in turn adversely affects hunger and appetite. One of the first studies to show the impact of sleep deprivation on eating patterns came from the University of Chicago in 2004.[155] When people slept just four hours a night for two consecutive nights, they experienced a 24 percent increase in hunger and gravitated toward high-calorie treats, salty snacks, and starchy foods. This is probably due to the body's search for a quick energy fix in the form of carbs, which are all too easy to find in the processed, refined varieties. I also get very hungry, particularly for carbs, when I stay up late writing grant applications.

Dr. Paolo Sassone-Corsi is a professor of biological chemistry and director of the Center for Epigenetics and Metabolism at the University of California–Irvine. Paolo has been studying circadian rhythms for more than twenty years, in particular the relationship between circadian rhythms, timing of eating, and weight control. Calories cannot tell time, but the body can through its circadian rhythms. He has taken genetically identical mice and fed them exactly the same food, but at different times. One group was fed at the normal time for its internal clock, while the other was fed at the wrong time. The normal-time group remained normal, while the other group got fat. One interpretation is that the stress induced by eating food at a time incompatible with one's metabolic cycles changes how the body handles that food. But see the next section, too.

Paolo has been a key player in our understanding of how hormones are linked with our circadian clock and how proteins involved with circadian rhythms and metabolism are intrinsically linked and co-dependent. For example, the *Clock* gene produces a protein, CLOCK, that is an essential molecular gear of the circadian machinery. CLOCK interacts with a protein called sirtuin 1 (SIRT1), which senses cell energy levels and regulates metabolism and aging. When these important interactions are disrupted or off balance, normal cellular function can be compromised, leading to illness and disease. This can also be related to sleep patterns and diet, because quality sleep and nutrition may help maintain or rebuild the balance and also explain, at least in part, why sleep deprivation or disruption of normal sleep patterns can increase hunger, leading to obesity-related illnesses and accelerated aging.

In 2015, the National Sleep Foundation, along with a group of experts, issued its new general recommendations for sleep. Babies, for instance, require more sleep than elderly individuals. But bear in mind that the recommendations are mostly generated by averaging how much we slept historically. There are very few studies that can say precisely how much sleep each of us needs. Individual sleep needs will probably be somewhat different, and the most important sleep number to maintain is the right amount of the restorative, slow-wave deep sleep. Although we need just one to two hours of slow-wave sleep per night, we know little about how to induce or facilitate it.

Millions of Americans suffer from sleep disorders, the most common ones being insomnia and sleep apnea. An array of factors can cause insomnia, from medications and medical conditions to alcohol and caffeine consumed too late in the day. While drinking can make you sleepy, it also disrupts restorative, slow-wave sleep. Sleep apnea affects a whopping twenty-two million of us and is caused by a collapse of the airway during sleep; muscles in the back of the throat fail to keep the airway open. This results in frequent cessation of breathing, which causes sleep to be fragmented. Dreamless sleep and loud snoring are telltale signs of sleep apnea. Sleep apnea has a strong relationship with obesity; in fact, the most common cause of sleep apnea ***is*** obesity. That extra weight and fat around the neck help to trigger the airway collapse. Although sleep apnea can be treated, usually with the help of a continuous positive airway pressure (CPAP) device, the best solution is weight loss. People who lose weight often find relief and may no longer need a CPAP device.

The fact that shift work has been linked to obesity, heart attack, several types of cancer (breast, prostate), a higher rate of early death, and even lower brainpower has everything to do with the connection between light and our circadian rhythms. People who work night shifts may think they can "train" their bodies to happily be up at night and sleep during the day, but the research tells a different story.

Documenting this association started when two of Dr. Eva Schernhammer's colleagues developed cancer with no risk factors or a history of the disease. These were healthy women in their thirties. At the time, Schernhammer worked rotating night shifts in a Vienna, Austria, cancer ward from 1992 to 1999, where in addition to her regular hours she had to work through the night ten times a month. When she moved to Harvard Medical School three years later, Schernhammer looked at the available medical, work, and lifestyle records of nearly seventy-nine thousand nurses enrolled in the famous Nurses' Health Study. Remarkably, she discovered that those who had worked thirty or more years on night shifts had a 36 percent higher rate of breast cancer, compared with those who had worked only day shifts.[156] She continued to investigate further and documented more disturbing findings. By late 2005, she published reports that her fellow female night owls had a 48 percent rise in breast cancer. In comparison, blind women had a 50 percent ***reduced*** risk of breast cancer. Multiple studies have since confirmed links between shift work and higher risk of a variety of cancers and cardiovascular and metabolic disease.[157] The causes most often given are the repercussions of a disrupted circadian rhythm and low levels of the sleep-inducing hormone, melatonin. Very recent research has linked these to changes in the epigenome.[158]

WHEN WE EAT IS AT LEAST AS IMPORTANT AS WHAT WE EAT

Satchidananda Panda is a professor in the Regulatory Biology Laboratory at the Salk Institute for Biological Studies in La Jolla, California. He has worked extensively on the circadian clock, especially as it relates with our genes, microbiome, eating patterns, risk for weight gain, and even our immune system. One of his most important discoveries showed how light sensors in the eyes work to keep the rest of the body on schedule. The hypothalamus is the part of our brains that links the nervous and endocrine systems—it regulates many of the autonomic functions of our body, particularly metabolism. The suprachiasmatic nuclei in the hypothalamus receive input directly from these light sensors in the retinas of our eyes and serve to "reset" our circadian clock. This is why exposure to early morning light helps reset the circadian clock and explains why getting out into the morning sun can help to recalibrate your clock after a late night or from jet lag.

No doubt you have read in many diet books that it is best to eat small meals throughout the day rather than three larger meals. This is an example of something that seems to make sense (frequent small meals are thought to prevent hunger pangs) but that actual experiments showed to be completely false. Strikingly, Panda and his colleagues found that mice consuming their calories within a set amount of time (twelve hours) were slimmer and healthier than those that ate the same number of calories but over a larger time window.[159] Put another way, the periodic fasting between meals, followed by a longer period of fasting during the sleep period, made mice leaner and healthier than mice that consumed the same number of calories throughout twenty-four hours. But why? We don't fully understand how a time-based eating pattern can prevent weight gain and illness, but Panda and his colleagues think that the timing of meals influences the circadian clock, which in turn affects the function of genes that are involved with metabolism.

The microbiome also has diurnal cycles that play an important role here. Panda has done extensive work on how the gut bacteria wax and wane throughout the day and how such variations hinge on day and night cues. In one very important study, he and his colleagues compared the microbiomes of mice fed normal food with those given high-fat fare. They measured the composition of the gut microbiome every four hours, rather than taking a snapshot once a day. In the mice on normal diets (that ate during the night and slept during the day), Panda documented dramatic fluctuations in the particular genera of bacteria present at any given time. In the mice that ate a high-fat diet and generally fed around the clock, however, not only did their microbiomes not fluctuate, but these mice gained weight and developed diabetes.[160] These patterns were evident among multiple species of bacteria, including Firmicutes, a type of bacteria that has been associated with obesity and disease in multiple studies around the world. Panda's group was able to show that it is not necessarily whether an organism has high or low Firmicute levels that dictates health, but when or how often those levels peak. Healthy mice can have high Firmicute levels during the night when they normally eat, but this naturally wanes during the day when they should be sleeping. However, in obese mice, Firmicute levels remain high all the time.

If the thought of timing your meals more precisely throughout your day is starting to stress you out, then I invite you to read on.

STRESS

We all know that chronic, unrelenting stress does not do a body good, and many of us are under unprecedented levels of stress today. According to an American Psychological Association survey, about one-fourth of Americans rate their stress level as 8 or more on a 10-point scale.[161] In general, each individual responds to stress a little bit differently. Some of us turn to comfort foods to soothe our anxiety, while others simply cannot eat. When the body is under high levels of acute stress—say, before you have to take an important exam or need to do something you dread—high levels of adrenaline in the body will suppress appetite in the short term. Your brain sends signals to the adrenal glands atop the kidneys to pump out the hormone epinephrine (adrenaline), which shuts down your hunger cues, while your hypothalamus produces corticotropin-releasing hormone to increase secretion of the stress hormone, cortisol, thereby preparing the body to recover from stress. This primes you to engage in a fight-or-flight response to the stress and then to recover from it.

But what happens when stress becomes chronic, when it does

not disappear? Austro-Hungarian endocrinologist Hans Selye is generally credited with coining the term "stress," or "general adaptation syndrome."[162] Selye observed that chronic stresses of different kinds induced more or less the same response, swelling of the adrenal cortex, increased liver weight, atrophy of the thymus, and ulcers. He recognized that this was a pathological response to the constant presence of a stressor. This type of stress, such as dealing with everyday demands at work and home, provokes a different response in the body that can include what some call "emotional eating"—overeating indulgent (sugary, fatty) foods. The stress hormone, cortisol, is intended to increase appetite and stimulate recovery from acute stress. The combination of high cortisol, high insulin due to food intake, and high motivation to eat is a recipe for weight gain and obesity. Cortisol not only tinkers with hunger cues, it tells your body to store more fat and break down tissues that can be used for quick forms of energy, including muscle. Prolonged high levels of cortisol can lead to increased abdominal fat (particularly that nasty, visceral kind), mood disorders (such as depression), bone loss, a suppressed immune system, fatigue, and an increased risk for insulin resistance, diabetes, and heart disease.

It is important to note that the brain's reward system, which regulates our ability to feel pleasure, is also at play. The addiction literature has long documented how the reward system is impacted by drugs such as nicotine, cocaine, and heroin—powerful substances that target the pleasure center and keep addicts coming back for more. This same literature now suggests that this reward circuitry may have a central role in stress-induced food intake. So similar to drug addicts, who were once labeled as people who lacked self-control, food addicts can partly blame their "lack of control" on how their brain is speaking to them and creating impulses to eat. Did you ever see the "Betcha can't eat just one" commercial? Stress, together with highly palatable food, can stimulate the release of "intoxicating," feel-good brain chemicals—natural opioids, in fact—that keep one coming back for more.[163] The release of these opioids may actually be part of our bodily defense mechanism against the stress response, helping us to turn its volume down. But repeated stimulation of the reward pathways through the stress response, coupled with an intake of highly tasty food, can lead to neurobiological adaptations that contribute to the compulsive nature of overeating.[164,165]

Overeating is not the only stress-related behavior linked to weight gain. Stress makes you less likely to exercise, more likely to turn to comfort foods or to substances such as alcohol, and more likely to have trouble sleeping—all of which further contribute to excess poundage.

PHARMACEUTICALS

A variety of prescription drugs we take have the unpleasant side effect of causing weight gain. If you use any pharmaceuticals on a regular basis, read the fine print in their packaging to see whether "weight gain" or "changes in appetite or weight" are listed as side effects. (RxList and WebMD are good online sites to check for information.) This is not to say you should immediately discontinue their use, but raise the issue with your doctor. You should always be made aware of the expected benefits as well as potential risks of any medical intervention, particularly prescription drugs.

The fact certain drugs can cause weight gain is perhaps the

strongest proof that chemical obesogens exist and can cause obesity in humans. Drugs are merely chemicals tested for effectiveness against a particular condition. Many of these drugs target the same physiological pathways as chemical obesogens (for example, tributyltin and the diabetes drug rosiglitazone, aka Avandia). Some medications can make you feel hungrier, while others slow your ability to burn calories or cause you to hold on to extra fluids. However, not everyone responds equally to the same drugs. You might not gain an extra pound while taking a certain drug, while your friend puts on fifteen pounds on the same medication. Again, this is evidence as to the individuality of human biology. Many factors come into play, from underlying genetics and epigenetic programming to environmental impacts and exposures. Let's take a quick tour of the most common pharmaceutical obesogens.

Depression medications (for example, Celexa, Prozac, Paxil, Zoloft): The so-called SSRIs (selective serotonin reuptake inhibitors) as well as "tricyclic antidepressants" have a long history of being associated with weight gain.[166] The tricyclic drug amitriptyline (Elavil) in particular has been shown to induce sugar cravings and considerable weight gain in some people.[167] Keep in mind that depression itself can affect your appetite and eating habits as well. Note that some of these medicines, such as Elavil and Pamelor, are also used to treat seizures and migraines.

Mood stabilizers (such as Clozaril, Seroquel, Risperdal): These drugs help treat mental health conditions such as bipolar disorder or schizophrenia. Through their effects on the brain, they affect your weight and metabolism by keeping your appetite turned on. Some may cause as much as an eleven-pound weight gain in ten weeks. People taking them for a long time may gain more. The antipsychotic drug olanzapine (Zyprexa) in particular has been shown to induce considerable weight gain and obesity.[168,169]

Diabetes drugs (for instance, Diabinese, Insulase, Actos, Avandia): These drugs control blood sugar levels, but do so in different ways. Some make you more sensitive to insulin, while others trigger the body to release more insulin before or after meals. Weight gain is normal when starting these drugs, for the body takes time to adjust. But older generations of these drugs such as Actos and Avandia can lead to substantial increases in body fat content.[170] People with type 2 diabetes who are already overweight find the extra weight particularly frustrating.

Corticosteroids (such as Medrol, prednisone): Corticosteroids reduce pain and inflammation and are not the same as the anabolic steroids often abused by bodybuilders and athletes. Corticosteroids have long been known to trigger appetite and foster fat storage in humans.[171] They are often prescribed to treat an array of different conditions, such as severe allergies or skin problems, asthma, arthritis, or Crohn's disease.

Over-the-counter allergy medications (for instance, Zyrtec, Benadryl, Allegra, Claritin): Over-the-counter allergy meds block the action of histamine, a chemical your body makes that causes many of the symptoms of allergies. A 2010 study published in the journal *Obesity* reported an association between the use of antihistamines and obesity. The study found that of nearly nine hundred people studied, those taking antihistamines were more likely to be overweight or obese than those not taking these drugs.[172] The underlying reasons for this were not clear, and this association does not prove that the antihistamines caused obesity. The researchers speculated that antihistamines have similar chemical structures to certain psychiatric drugs that are known to be associated with weight gain, as they may increase appetite.

VIRUSES

Interestingly, evidence is accumulating that exposure to some viruses, such as the adeno-associated virus 36, may influence risk for becoming overweight or obese. In the past several years, a number of studies in humans have shown that infection with adenovirus 36, a virus that causes upper-respiratory infections, leads to fat weight gain, especially in children.[173] These studies support earlier research that showed the virus causes weight gain in mice, rats, chickens, and monkeys. How this happens is similar to how some chemical obesogens work: by prompting adult stem cells in fat tissue to make more fat cells, which then store more fat. When researchers examined the fat pads from infected animals and compared these with those of uninfected animals, they noted that the infected animals had bigger fat cells and more of them. When they infected human adult mesenchymal stem cells in tissue culture, the cells differentiated into fat cells—***without even adding differentiation agents to the culture***. In 2013, a team of researchers at Pennington Biomedical Research Center at Louisiana State University published a study that followed fourteen hundred people, finding that those who tested positive for antibodies to the virus—indicating that they had been infected at some point—gained significantly more weight in the form of body fat over a ten-year period than those who had not been infected.[174]

Counterintuitively, despite its effects on obesity, exposure to this virus seems to improve other metabolic parameters, notably lowering cholesterol and triglycerides and improving blood sugar control. But the researchers studying this virus do not yet know why and are currently searching for ways to manipulate the virus to promote such positive outcomes while minimizing the fat-inducing effects. In addition, not every obese individual has these exposures, and not all of those who have been exposed are overweight or obese.

FAT GENES

I would be remiss to leave genetics out of this discussion. Are some people saddled with "bad genes" that make them more susceptible to weight gain and sensitive to chemical exposures? The answer is yes, but not so many. That is, there are genes that have been associated with obesity, but the existence of such genes cannot come close to explaining the magnitude of the obesity epidemic, as we discussed already. The most recent, large-scale meta-analysis of genes that might be associated with obesity identified ninety-seven regions of the genome that were associated with variations in BMI. This analysis included 339,224 individuals from 125 previous studies and concluded that these ninety-seven regions could account for about 2.7 percent of the variation in BMI among individuals—a very small amount indeed.[25]

Some of the genes identified were linked with the central nervous system (that is, the regulation of eating and metabolism), whereas others were associated with expected factors such as insulin signaling, the development of fat cells, the regulation of fat import/export into fat cells, and overall energy metabolism. The researchers proposed that these results indicated that as much as 20 percent of variability in BMI might be explained genetically—still a very small number. The bottom line is that genetics does play a role in susceptibility to obesity, but that probably 80 percent or more of obesity has causes other than mutations in genes that have been, or can be, identified. This is not so different from the overall association between genes and diseases of about 20 percent.

PART II

In the Real World: How to Take Control in 3 Simple Steps

CHAPTER 6

Step 1: It Really Does Start with Food

Cleaning Up Your Diet

It is the number one question I get from people who hear me speak about obesogens: *What is the first thing I should do today?* And my answer is always the same: Clean up your diet. This is easier than you think. You do not have to move to the wilderness and grow your own food off the grid. Nor do I mean that you should launch yourself into a fad diet designed for weight loss, start juicing exclusively, or become vegetarian or vegan. Cleaning up your diet means following a few basic guidelines that I will give you in this chapter. I call them the "Anti-obesogen Rules." They will reduce your burden of obesogenic chemicals and other EDCs, while helping you to avoid additional exposures. And as an automatic side benefit, you will be supporting local farmers who produce food responsibly and sustainably.

We live in a world of choices now when it comes to our food, but we also live in a world of confusion as a result of food industry tricks that make us think we are eating "healthy" meals. Even the most educated consumers can fail a test about the definition of natural, organic, local, gluten-free, and grass-fed. Journalist Kristin Lawless has written an excellent book on this topic entitled *Formerly Known as Food*.[175]

RULE #1: THINK WHOLE, FRESH, AND UNPROCESSED

Make it your main goal to prepare as many of your meals and snacks as possible yourself, using whole, fresh, unprocessed ingredients that you select (therefore, you will know precisely what is in your food). I try to make all my meals at home. I'm in charge of the grill, and my wife is a great cook who makes everything from fresh ingredients. Together we can do better than any food bar in a market and any restaurant. We do better in food quality and flavor than all but perhaps the highest-end gourmet restaurants. When you do venture outside your kitchen, patronize places that use fresh, organic, and locally grown ingredients. Even better would be where you can ask the chef questions about the food. The "farm to table" movement is growing everywhere across the country and is no longer confined to places such as my state of California. Look for those types of restaurants in your area. In general, top-quality restaurants do have better standards and suppliers than the alternatives. But beware: a 2016 study done at the University of Illinois showed that when you go out to eat, whether at a full-service or fast-food restaurant, you will consume an average of two hundred more calories, 10 grams more of fat, and twice the sodium than if you had eaten a meal prepared at home.[176]

If you buy preprepared foods at markets, which should be minimized, look for fresh ingredients devoid of artificial preservatives, colors, flavors, sweeteners, and hydrogenated fats. Keep in mind that if you didn't make it, you simply do not know what is in the food you buy.

You have probably already heard about the benefits of "shopping the perimeter" of a store. This is where most of the fresh (meats, produce) and least processed (dairy) foods are located. The inner aisles are where the foods sold in boxes, bags, and cans are found. If you buy anything that comes with a nutrition label (note that most fresh items such as produce, fish, and meats do not), become skilled at reading those labels. When she was little, my daughter used to ask me, "Papa, how do you know that this food is not good?" My answer was simple: "Do you put that many ingredients in foods you make with your mom?" Long lists of ingredients automatically tell you that the food is highly processed and probably not nearly as good as food you make yourself from fresh ingredients. Words you cannot pronounce or define are probably chemicals you should avoid. In addition to avoiding chemical obesogens, you want to stay away from ingredients that qualify as nutritional obesogens:

Added and processed sugars (for example, high-fructose corn syrup). Stick with foods that do not list sugar at the top of their ingredients. Better yet, avoid foods with added sugar at all. Honey and agave syrup contain high amounts of free fructose—avoid them unless you are a diabetic who cannot tolerate sucrose.

Artificial sweeteners (for instance, acesulfame potassium [Sunett, Sweet One], aspartame [NutraSweet, Equal], saccharin [Sweet'N Low, Sweet Twin, Sugar Twin], sucralose [Splenda], and neotame [Newtame]). Remember, recent studies show that many artificial sweeteners can damage your metabolism by altering the composition of your microbiome.[19] They make you much more susceptible to overeating and can also trigger transient increases in insulin levels (increasing storage of fat). Food companies are now responding to the increasing public concern by using obscure names to hide these synthetic sweeteners in their products. The list of artificial sweeteners is long and continues to grow with new formulations. They not only lurk in many prepared foods such as salad dressings, baked goods, processed snack foods, "lite" and diet foods, and breakfast cereals, but can be found in unsuspected places such as toothpaste, liquid medicines, chewing gum, and frozen desserts. Note also that the jury is still out on sugar alcohols such as sorbitol, mannitol, xylitol, maltitol, erythritol, and isomalt. While these sugars may not produce as significant a rise in blood sugar as other sugars (and are often marketed as healthier alternatives to regular and artificial sugars), not enough studies have been performed on them. For example, we don't know anything at all about the effects of sugar alcohols on the microbiome and, in turn, metabolism.

Monosodium glutamate (MSG). This flavor enhancer is famously linked with "Chinese restaurant syndrome" (although this might not be completely true), but it is also found in a variety of processed, packaged foods such as chips, frozen dinners, cold cuts, dressings, and salty-flavored snacks. Food manufacturers know that you are avoiding MSG, so they label it in the ingredients with euphemisms such as hydrolyzed protein, glutamic acid, yeast extract, or autolyzed yeast.

Other additives (for example, nitrites and nitrates, potassium bromate, artificial colors). There are more than ten thousand additives allowed in food, most of which are classified as "generally recognized as safe," or GRAS.[x] Why is this bad? To qualify for the GRAS designation, it is only required that there is a substantial history of consumption for food use by a "significant number of consumers" (whatever that means).[177] In nearly all cases, there is little or no testing to support this designation—only a substantial history of use. I recommend referring to the website of the Environmental Working Group (EWG)[178] for their list of top additives to avoid. You can also look up your food in their database for more information. Not all additives are obesogens per se, but avoiding them will help you go a long way toward cleaning up your diet in general and steering clear of EDCs and potential obesogens that have not been identified yet.

It is probably unrealistic in the modern world to avoid ***all*** packaged foods, so do the best you can. For example, when you purchase condiments and acidic foods such as peeled tomatoes, aim to buy them in glass rather than plastic. Minimize canned

---

[x] Under sections 201(s) and 409 of the Federal Food, Drug, and Cosmetic Act, any substance that is intentionally added to food is a food additive and subject to premarket review and approval by the FDA, unless the substance is generally recognized, among qualified experts, as having been adequately shown to be safe under the conditions of its intended use, or unless the use of the substance is otherwise excepted from the definition of a food additive. The term "qualified experts" is open to interpretation, and in my opinion, GRAS should actually be interpreted as "generally recognized as untested, but perhaps safe."

goods overall. Store your own foods in glass or stainless steel so that you eliminate exposure to chemical obesogens found in storage containers that can leach into the clean fresh food you worked so hard to make (more on this in chapter 7).

RULE #2: BUY ORGANIC, GRASS-FED, AND WILD WHENEVER POSSIBLE

Buying organic has become much easier and more economical in recent years, and choices have increased greatly. You once had to shop at gourmet specialty stores, a few national retailers who charged very high prices (you know who they are), and farmers' markets to find organic foods. Happily, the increased demand has compelled mainstream grocery chains and big-box stores such as Costco and Walmart to sell certified organic foods. What does it mean to be certified? The USDA Organic Seal indicates that a food was produced without synthetic pesticides, genetically modified organisms (GMOs), or fertilizers made from petroleum. When it comes to organic meats and dairy products, the seal bears even more weight: it also means that the meats or dairy products are from animals that are fed organic feed and forage, are not treated with antibiotics or hormones, and are raised in living conditions that accommodate their natural behaviors, such as grazing.[179]

Let's take the bull by the horns and talk about the costs vs. the benefits of organics. There are a variety of recent studies that purport to test whether or not organic produce is more or less nutritious than conventionally grown.[180-182] Many of these miss the point nearly completely. While we can argue endlessly about whether or not one batch of organic spinach has more or less nutrients than a batch of conventionally grown spinach (and organic probably has at least a bit more), it is ***indisputable*** that the organic spinach contains little or no potentially toxic or endocrine disrupting agrochemicals (pesticides, herbicides, fungicides, and fertilizers). The point of eating organic is that you dramatically reduce your exposure to these chemicals, not that the food is necessarily more nutritious. At the same time, organic agricultural practices (crop rotation, tilling the soil, plowing leftover vegetation back into the soil) improve the quality and nutrition of the food as well as the health of the soil and environment. Patrick Holden in the United Kingdom has been leading the charge for wholesome, sustainable food production. Visit the Sustainable Food Trust website[183] to learn more about how important this is.

Organic labeling can also be misleading. Only foods made with 100 percent organic ingredients can indicate that on the label. If a food was made with at least 95 percent organic ingredients, then it can say just the word "organic." Products that say "Made with organic ingredients" were created with a minimum of 70 percent organic ingredients, with restrictions on the remaining 30 percent, including no GMOs. This is why I recommend trying to purchase foods that have the "USDA Organic" or "100% certified organic" label. I also suggest that you try to buy organic products grown in the United States and avoid those produced in foreign countries, even if they are sold with the "100% certified organic" seal. Not only has the *Washington Post*[184] reported on large imports of conventionally grown food that were improperly labeled as organic, but a newer audit[185,186] conducted by the USDA's Office of the Inspector General has confirmed serious concerns about imported foods labeled as "organic" but are anything but. Until all the loopholes and weak links are dealt with in the global supply chain, it behooves us all to be extra skeptical of "organic" goods produced on foreign soil and sold in the United States.

You can also find foods produced in the United States that are sold as organic but do not carry these official certification labels. I would be suspicious about these. Although there may be nothing wrong with these foods, they clearly do not meet the standards to be labeled "USDA Organic." This might be because the farmer does not have the resources to have his farm certified, or it could be because the farmer is using organic practices, but insufficient time has passed for the farm to be certified. The lack of USDA certification means that you must know where these foods came from and how they were made or farmed in order to trust them (see sidebar). To this end, shop at reputable markets (often larger regional or national chains) where you can trust the origins of your foods and feel comfortable asking questions at the butcher and fish counter or in the produce section. While I don't mean to denigrate small, independent markets, large chains have more to lose by cutting corners. On the other hand, small, independent markets can offer personalized services that the chains do not. They may also offer fresh foods that are grown more locally and with minimal time spent being transported.

QUESTIONS TO ASK YOUR LOCAL FARMER

If you have a farmers' market near you, you will likely run into merchants selling goods they claim are "organic" and/or "locally grown" but that do not have official certification labels on them. Some of these foods are in fact fine to eat (and might meet organic-certifiable standards). Here are some questions you can ask:

- Where is your farm located?
- Are your foods 100 percent certified organic? Certified by whom?
- How is this [fruit/vegetable] grown?
- How do you control for insects and weeds? Do you use pesticides or herbicides?

Trustworthy organic farmers tend to be proud of their practices. They will openly engage with you and share their farming methods. If you are buying foods that come from animals, such as cows or chickens, ask what they feed them. And if at any time you feel you are not getting clear answers when you ask questions, then move on to the next booth.

My family has been in the food industry for generations. My grandparents on both sides of the family owned competing grocery stores on South Street in Philadelphia (my parents' marrying was a sort of Romeo and Juliet love story with a mostly happy ending). Both of my grandfathers were butchers, as was my father. I worked in the seafood department of a local supermarket in high school to pay for my car and to take my high-maintenance girlfriend on dates. When I turned eighteen, I served a three-year apprenticeship and became a journeyman butcher, which is how I worked my way through college and later financed my relocation to California before entering graduate school in 1982. As you can imagine, I got to see what goes on behind the scenes, and it was not a pretty picture. Remember the old adage (misattributed to Otto von Bismarck) about how we should not know how laws or sausages are made? So true, but

I am not so certain that lawmaking is cleaner than sausage making today.

I can tell you that in the meat departments of many markets I worked in over the years, large chain, small chain, and independents (more than one hundred overall), it was commonplace to put a "prime" sticker on the prettier steaks for a higher price, even when they were not prime. Ground sirloin? Haha, that's a good one. This was ground meat with a higher proportion of tasteless imported beef, colloquially called "bull meat" to give it the stated fat content. At more than one market I worked at in New Jersey, the butchers were instructed to spray something on steaks and roasts that had gone off-color to restore them to red. We called it "dynamite" because it blew away the green color. This was known to be illegal, but few stores were ever caught in the act. At least a few stores I knew of used frozen chicken giblets to make their hamburger bright red. Some small, independent markets I worked at sold their off-color steaks to local restaurants at a discount (no doubt supporting a few "kitchen nightmares"). Shenanigans in the meat and seafood industry are widespread. Now there is a high-tech mechanism to identify the species in food known as the FoodExpert-ID DNA chip.[187] This is not something that you can buy, or even that your local board of health can afford, but it is at least slightly comforting to know that it exists.

I am not saying that this sort of behavior is common today, but knowing what I know, I would not exclude the possibility either. Not all markets hold themselves to high standards. When I go shopping, I most frequently purchase meat and seafood at Costco, which sells in very high volumes—selling in great quantities quickly so nothing sticks around long enough to get old—and the employees have little time or motivation for shenanigans. Other good choices might be a high-volume store in a regional or national chain. If you know of a reputable meat market or fish shop, by all means support it. Sadly, there is a reason why we hear all-too-frequent reports of food contamination in the news—corners are being cut. In my day, every wholesaler had an on-site USDA inspector who had the power to shut the entire place down in an instant if a violation was observed. I don't know exactly when that stopped (I left the industry in 1982); now there are only a few USDA inspectors per city to monitor what happens at the wholesalers.

Here are some quick tips to protect yourself:

- Meat that did not sell fast enough and got a little green will be ground up first thing in the morning—we called them "rewraps." Needless to say, do not buy ground meat first thing in the morning, particularly the very lean, higher-priced stuff—it is almost guaranteed to have the less fresh meat in it.
- Steaks and chops might be seasoned to sell quickly at a discount. Avoid pre-seasoned meats, particularly when the seasoning covers the surface so you can't see whether it is off-color (in other words, old) or not. I never once saw fresh meat seasoned like this before sale—it was always the off-color stuff.
- Trust your nose. Fresh meat and fish have only the faintest of smells.
- I myself would never, never, not ever buy pre-ground meat from any market, anywhere. Select a steak, roast, or package of stew meat and ask the butcher to grind it up for you. And watch him do it so that he does not substitute something else. Better still, buy a KitchenAid or similar

multifunction mixer with a grinder attachment and make your own ground beef, pork, chicken, turkey, and so on.

When cattle are fed grains, usually a feedlot made with corn and soybeans, the quality of their meat products will be different from what they would be had the cattle been fed their normal diet of grass and hay. Grass-fed cows are raised exclusively on hay or grass after they are weaned, and the beef produced from them is leaner than conventional beef, yet their fat tends to have a higher proportion of omega-3 fatty acids (the good fat). But be careful: ask whether the meat is 100 percent grass-fed. It is commonplace to graze the animals for most of their lives and then "finish" them in high-density feedlots where they are crowded, exposed to diseases, given sub-therapeutic levels of antibiotics, and fed with cheap, GMO corn (see page 174) to quickly fatten them up for market.

Grass-fed does not imply organic, either. Grass-fed cattle can still be pumped with hormones and antibiotics, so look for grass-fed organic meats. These do exist, and the meat can be excellent. I have eaten many a beautifully marbled, grass-fed steak in Buenos Aires, Argentina. Beware of markets selling organic meats that have not been USDA graded (for instance, USDA Choice or USDA Prime). My wife spent almost $30 a pound on some grass-fed New York steaks sold by a famous national retailer that were chewy, gamy, and mostly inedible. The meat may have been free of hormones and additives (hence the organic labeling), but that does not mean it was high quality with respect to taste, texture, and tenderness. It wasn't. I returned the uneaten bits to the market and told the meat manager that these steaks he was selling at an extraordinarily high price were, in my expert opinion, not even up to pet-food-quality standards. He was not amused, but neither was I.

When buying eggs, you will sometimes see "pastured eggs" in addition to "organic eggs." The difference? Organic eggs come from chickens that are not treated with antibiotics or hormones and that eat organic feed. They also have limited access to the outdoors. Pastured eggs, on the other hand, are similarly not treated with antibiotics or hormones but are produced by hens allowed to roam more freely and eat their natural food, plants, and insects. Pastured eggs tend to cost a lot more than organic, and it's fine to stick with the organic eggs.

When buying fish, try to avoid farmed fish and go wild; choose to eat smaller, presumably younger fish. Why do I say this? Everyone knows that commercial fishing can be very destructive to the environment and to wild fish stocks. Unfortunately, fish farming, as commonly practiced, is even more destructive to the environment, although it can preserve wild stocks. Since this is a book about our health, I must urge you to protect your own health first. Farmed fish are often fed unnatural diets that lead them to be contaminated with antibiotics, PCBs, and other toxic chemicals. Wild salmon typically eat small crustaceans and zooplankton. Farmed salmon are fed a food based on soy and fish meal and need added coloring to attain their normal color. Try to avoid (or at least limit your consumption of) fish high on the food chain, such as shark, swordfish, pike, albacore, bluefin tuna, halibut, king mackerel, and tilefish. Persistent pollutants such as DDT, PCBs, and mercury are bioaccumulated up the food chain, leading to high levels in top predators. Examples of fish low on the chain that can be consumed more liberally include wild Pacific salmon, pollock, anchovies, sardines, herring, sablefish/black cod, and sole. Avoid freshwater fish from the Great Lakes and other polluted, industrialized areas. I avoid buying fish imported from places that I know or suspect have lax environmental and food-quality standards.

The best way to know whether or not you're buying high-quality fish is to inspect the fish yourself. Be discerning. Whole fish should have clear eyes. Fish should never have a strong odor. Anything more than a very slight smell indicates that the fish is probably at least four or five days old. You also want to be sure that you are getting what you intend to buy. Mislabeling is a pervasive problem today in both stores and restaurants. At many California markets I worked in, we received a common local fish called "rock cod." With the skin on, this was sold as "red snapper"; with the skin side down or off it was "ocean perch"; and particularly thick pieces were labeled "sea bass." Numerous investigations in recent years have uncovered other unsettling bait-and-switch cases. Is that Chilean sea bass or Patagonian toothfish? Wild Pacific salmon or farmed Atlantic? Giant sea scallops or plugs of shark meat? Looks can be deceiving, so ask questions. This issue is more important than just paying too much for the wrong fish. There are health consequences to eating a contaminated, farmed fish when you expect to be eating a wild-caught fish free of antibiotics, hormones, toxic chemicals, or bacteria from poor husbandry.

The produce department also has many potential issues. When you cannot access or afford organic produce, at least make it your mission to avoid the "Dirty Dozen." These are fruits and vegetables that consistently have the highest levels of unhealthy pesticides, herbicides, and fungicides many of which are EDCs and could also be obesogenic. The Dirty Dozen is a list compiled and updated annually by the Environmental Working

Group upon analyzing pesticide residue testing data from the U.S. Department of Agriculture and the U.S. Food and Drug Administration.[188] The EWG also maintains a "Clean 15" list to indicate which produce in a particular year contains the lowest amount of pesticide residues and is therefore safer to buy "conventionally grown" if organic is not available.[189]

Go organic when you buy any of the 2017 "Dirty Dozen":	Instead pick from the 2017 "Clean 15":
1. Strawberries	1. Sweet corn*
2. Spinach	2. Avocados
3. Nectarines	3. Pineapples
4. Apples	4. Cabbage
5. Peaches	5. Onions
6. Pears	6. Sweet peas (frozen)
7. Cherries	7. Papayas*
8. Grapes	8. Asparagus
9. Celery	9. Mangoes
10. Tomatoes	10. Eggplant
11. Sweet bell peppers	11. Honeydew melon
12. Potatoes	12. Kiwi
	13. Cantaloupe
	14. Cauliflower
	15. Grapefruit

** A small amount of sweet corn, papaya, and summer squash sold in the United States is grown from genetically modified seeds. Buy organic varieties of these crops if you want to avoid genetically modified produce.*

If you want a fruit or vegetable listed among the Dirty Dozen and local, fresh organic is not available, I recommend an alternative organic selection, even if it is not local. Or consider organic flash-frozen produce, which can be more tasty and healthful than some of the fresh produce sold in grocery stores. Fruits and vegetables destined for freezing are typically processed at their peak ripeness when they are most nutrient packed. Unfortunately, these are packaged in plastic bags, which is not ideal, but it is better than fresh non-organic, in my opinion.

A Note About GMOs:

Although everyone seems to have an opinion on genetically modified organisms, commonly called GMOs, I will wager that relatively few people can define exactly what GMOs are and why they might be troublesome. GMOs are plants or animals that have been genetically engineered with DNA from other living things, including bacteria, viruses, plants, and animals. The genetic combinations that result do not happen naturally in the wild or in traditional crossbreeding. This is the heart of the controversy—that these are "unnatural." Let's dig into this issue for a moment.

GMO foods are often created to fight pests that can destroy crops, or to cultivate crops with certain desired characteristics. The reason a lot of Hawaiian papayas are GMO, for example, is that the ring spot virus decimated nearly half of the state's papaya crop in the 1990s. In 1998, scientists developed a genetically engineered version of the papaya called the Rainbow papaya, which is resistant to the virus. Now more than 70 percent of the papayas grown in Hawaii are GMO.

Many crops, in fact, have been engineered to create a product that is more robust or nutritious. A sweet potato grown across

Africa has been made to be resistant to a particular virus. GMO rice has more vitamins and iron. There are fruit and nut trees that are engineered to yield crops years earlier than they would normally. Bananas can even be genetically modified to produce human vaccines against diseases such as hepatitis B. So what is my position on GMOs and whether or not they are bad?

In my mind, I divide GMOs into three broad categories:

Category 1: A Category 1 GMO in my view is one that has been engineered to produce a nutrient that it would not otherwise contain. Golden rice is an example of a Category 1 GMO. It is rice that has been engineered to produce beta-carotene, a precursor of vitamin A. Vitamin A deficiency kills an estimated 670,000 children under the age of five each year in some parts of the world. The introduction of a fortified food easily grown to address this problem is, in my book, a good thing. I have no issue with this type of GMO.

Category 2: When a plant is engineered to be resistant to some pathogens, such as the case with Hawaiian papaya, I call this a Category 2 GMO crop. In principle, I am not totally against such crops because they provide a benefit, for example, by reducing or preventing crop loss due to viral infection without bringing along much risk. How effectively they do this is another matter. In addition, there are some cases of people being allergic to the product of the genes that have been added. Another issue is that the traits that have been genetically engineered into these crops spread to other crops, including those in a farmer's nearby organic fields, which is an unintended but largely unavoidable and unacceptable consequence.

Last, the use of such pathogen-resistant crops can lead to monocultures, where a huge portion of a particular crop is genetically identical. As has recently been found in India, the use of a GMO cotton resistant to one type of pest did not prevent the crop from being devastated by other pests and produced a much lower yield than had been promised by the developer. For all of these reasons, some degree of caution is indicated in the use of such GMO crops.

Category 3: Crops that are engineered to be resistant to one or more chemical pesticides and weed killers (such as glyphosate) are Category 3 GMOs. In my view these are the most concerning, because the added chemical resistance leads to (1) more chemicals being used, which leads to the development of resistance and increased levels of the chemicals in the environment; and (2) higher levels of potentially toxic chemicals on the GMO product. Big agribusiness is heavily promoting such pesticide-resistant crops for a variety of self-serving reasons, mostly concerned with increased profits, which I think is detrimental to both environmental and human health. Perhaps you have followed the recent controversy about the widely used herbicide glyphosate being carcinogenic? If not, you may want to read Carey Gillam's book, *Whitewash: The Story of a Weed Killer, Cancer, and the Corruption of Science.*[190] As far as we know at the moment, none of these chemicals are obesogenic, but there are quite a number of other adverse health consequences that research has associated with them. I am strongly against the use of Category 3 GMOs.

Current farming practices to grow GMO foods are another reason I am strongly against Category 3 GMOs. The image of farm workers yanking out weeds from the fields by hand is reflective of a bygone era. Now farmers spray the powerful weed-killing chemical glyphosate on their crops that are resistant to it, hoping that the crops outgrow the weeds. This saves a lot of work, at the risk of increasing environmental exposure to a chemical that research has associated with adverse health consequences, including cancer.[191] A burgeoning use for glyphosate is to kill and dry crops such as wheat so that they can be harvested when it is convenient for the farmers.[192] Glyphosate is again applied to kill weeds and make it unnecessary to till the soil in preparation for a new crop—so-called no-till agriculture. No-till agriculture is beneficial in the sense that less energy is used to plant the crops and soil loss due to runoff is reduced, but at the expense of degrading soil health and promoting the development of herbicide-resistant weeds. These are detrimental to the long-term health of the soil and the environment.[191]

To protect crops from the herbicide, the seeds are genetically modified to be resistant to its effects. These GMO seeds and their crops allow farmers to use massive quantities of herbicides such as glyphosate. The increasing use of herbicides means that GMO and conventionally farmed crops are almost invariably contaminated with herbicides and other agrochemicals. Corn and soy are the top two GMO crops in the United States, and it has been estimated that GMOs are in as much as 80 percent of conventional processed foods. This is largely because GMO corn and soy are ingredients in those processed foods. Did you know that sugar beets provide half of all consumable sugar in America and that 95 percent of those sugar beets are grown using glyphosate-resistant GMO seeds?

Most GMO crops are not directly consumed by humans. GMO corn is used in animal feed and high-fructose corn syrup, and GMO soy makes soybean and vegetable oils, soy protein, and more heavily processed soy such as lecithin and flavorings that land in a trove of processed products. If you are a conventional meat, egg, or dairy eater, the animals those foods come from are eating lots of GMO corn and soy. Even strict vegetarians and vegans are unable to easily avoid GMOs because most of those veggie burgers and soy dogs are likely made with GMO soy. Certified USDA Organic is the only way to avoid exposure to these agrochemicals.

For the record, I have nothing against eating soy as long as it is unprocessed and organic. One exception: It is probably best to avoid giving a baby soy formula; not only does this contain GMO-derived soy, but soy contains phytoestrogens. Early life exposure to estrogenic chemicals can increase the risk of obesity and the development of hormone-related cancers later in life.

GMO ALERT

Common foods that are made with GMO ingredients or are derived from GMO-exposed animals:

- Premade soups
- Frozen meals
- Milk and soy formula
- Juice drinks and soda
- Cereals
- Vegetable oils
- Tofu
- Conventional meat and dairy

When you buy organic, grass-fed, and wild foods, you will automatically be lowering your risk for consuming GMOs and their associated agrochemicals without even needing to worry about labeling. Also keep in mind that "organic" does not automatically equate with "healthy." Many organic junk foods line the shelves in supermarkets today, including candy and baked goods that are anything but healthy and waistline friendly, despite being organic. When in doubt, scrutinize the ingredients. Look for anything suspicious. The same goes for the gluten-free industry. Many people have chosen to go on a gluten-free diet today, and the popularity of this diet has spawned an industry of products labeled "gluten-free." These products may be free of gluten, but that does not mean they are not filled with other obesogenic ingredients such as refined sugar and additives.

RULE #3: CLEAN UP YOUR WATER

Regardless of how much you love the taste of your tap water or what the glowing report from your water supplier says about the contents of your water, I recommend buying a household water filter, at least for all of your drinking and cooking purposes. This is because most of the chemicals we produce and use in industry and agriculture eventually make their way back into our drinking water. The water industry is wrestling with how to deal with and remove these "contaminants of emerging concern." These also include pharmaceuticals we take that ultimately end up in the toilet after they pass through our bodies. Most of these contaminants are present at low levels, but we are better off without them in our water in the first place. How can we accomplish this? Or should we trust the water suppliers to do this for us? Perhaps you recall what happened starting in 2014 with lead in the water in Flint, Michigan? Similar conditions exist in many other cities and municipalities around the country. Trust in yourself, rather than in the EPA or some local agency, to protect your family's health. This is even more important today as the Trump administration systematically dismantles the already small amounts of protections provided by EPA regulations.

There are a variety of water treatment technologies available today, from simple and inexpensive water filtration pitchers you fill manually, to under-the-sink filtration systems with storage tanks, to whole-house carbon filters that will filter all of the water coming into your home from its source. The latter is ideal, particularly if you subscribe to a service that changes the filters regularly, because you can then mostly trust the water used in your kitchen as well as the water in your bathrooms. Choose which filter technology best suits your circumstances and budget: whole-house carbon; individual carbon filters on water spigots, refrigerators, and the like; reverse osmosis filters in the kitchen. Each type of filter has its strengths and limitations, and one type does not accomplish all goals. You will want to avoid water pitchers with filters because of their plastic containers.

An entire book could be written on how to remove contaminants from your drinking water. The Environmental Working Group just developed its national "Tap Water Database," which can tell you something about what sorts of contaminants are in your water and what health effects they might cause.[193] In turn, this can give you some idea of which filter technology might be most useful for you. Generally speaking, carbon or activated charcoal filters do a reasonably good job of removing cancer-causing organic chemicals (such as trihalomethanes) from your water but do a poor job of removing minerals such as lead, arsenic, or hexavalent chromium, which can also cause cancer. Reverse osmosis filters do a better job of removing minerals from water than carbon alone but may not remove all of the organic chemicals unless multiple carbon prefilters are used.

Remember to include the cost of maintenance and filter changes in your calculations. Most areas have companies whose business it is to provide pure water solutions. Talk with several to discuss what contaminants you wish to remove considering how much water you use. In my opinion, a whole-house carbon filter coupled with a kitchen reverse osmosis drinking water system is an excellent choice. I keep coral reef aquariums and also have a reverse osmosis deionization system that provides the extraordinarily pure water such aquariums require.

Note that while reverse osmosis systems do an excellent job of removing a broad variety of contaminants, they may not be a good choice if you live in a drought-prone state such as California. These systems waste a lot of water in the purification process—they typically use three or four gallons of raw water to produce one gallon of filtered water. If you want to use such a filter, then it is probably only reasonable to use it in the kitchen for filtering your drinking water. If you are contemplating a reverse osmosis system and expect to use a lot of this water, consider getting a higher-pressure stacked filter with high-efficiency membranes, which reduces waste water production. Whichever type of filter you choose, follow the recommended maintenance procedures to ensure that it continues to perform optimally. If you are a renter, a whole-house carbon filter may be out of the question, but you can still install removable carbon filters on showers and faucets or a reverse osmosis filter in the kitchen.

As contaminants build up, carbon filters will become less effective and can release adsorbed chemicals back into your filtered water so change them regularly. If you choose not to install a whole-house water filtration system, then consider putting filters on your showerheads. Shower filtration systems are easy to find, inexpensive, and they eliminate your exposure to vaporized chemicals (that is, in the steam you breathe in the shower). If you really want to get on top of what is in your water, consult the EWG database, read the water reports from your local supplier, and consult with a water filtration company to understand what contaminants can be removed by which types of filters. If you have well water, it is definitely worth having the water tested at least yearly to monitor for the presence of contaminants of concern (particularly if you live in a farming region where agrochemicals such as atrazine or glyphosate may contaminate groundwater).

Drinking or using water from plastic bottles and containers should be avoided as much as you reasonably can. The longer water is stored in plastic bottles, the higher the concentration of potentially harmful chemicals, obesogens included. Bottled water samples have been shown to contain phthalates, mold, microbes, benzene, trihalomethanes, even arsenic. Moreover, bottled water is subject to approximately the same standards as the EPA drinking water standards, most of which have been set politically rather than scientifically.[xi] In addition to the problems associated with contaminants leaching from beverages stored in plastic, the plastic itself is an issue now; plastic pollutes our

[xi] At the federal level, bottled water must comply with the Federal Food, Drug, and Cosmetic Act (FFDCA) (21 U.S.C. §§ 301 et seq.) and several parts of Title 21 of the Code of Federal Regulations. Section 410 of FFDCA requires that the Food and Drug Administration's bottled water regulations be as stringent and as protective of the public health as the EPA's tap water standards.

environment and oceans the world over. According to the International Bottled Water Association, U.S. consumers purchased 11.7 billion gallons of bottled water in 2015, which translates to 88 billion half-liter bottles,[194] which is approximately equal to annual soft drink sales. Nearly all of these bottles are discarded and become litter, rather than being recycled. Consider this also: Toxins from degrading plastic containers will leach into watersheds, soil, and the oceans for a very long time to come. You can help stop this flood of chemicals continuing to enter our environment by removing plastic water containers from your life. If you want to tote a water bottle around, choose a durable, reusable container made from stainless steel, glass, or ceramic. I prefer stainless-steel vacuum bottles to keep my water cold and fresh.

RULE #4: AVOID SUGARY BEVERAGES

Sugar is a major source of our daily calories, and much of that sugar is coming in quickly guzzled liquid form from sweetened beverages. Every single day, half (five in ten) of adults and more than half (six in ten) of youths in the United States consume sugary drinks.[195,196] The average can of sugar-sweetened soda or fruit punch contains about 150 calories, almost all of them from sugar, usually high-fructose corn syrup. That is the equivalent of nearly ten teaspoons of table sugar ***in one beverage***. If you do the math, that means you could gain up to five pounds in a single year if you drank just one can of a sugar-sweetened soft drink every day without reducing caloric intake elsewhere in your diet.

Soft drinks do not include just your typical soda or carbonated, sweetened beverage. The term "soft drink" refers to any concoction with added sugar or other sweetener and includes soda, fruit punch, sweetened powdered drinks, sports and energy drinks, coffee, iced tea, lemonade, and other "ades." These drinks, even fruit juices, are nutritional obesogens. Aim to avoid or strictly minimize them—drink sparkling water from a glass bottle instead, perhaps with a squeeze of lemon, lime, or other flavors you enjoy. Consider sugary beverages as fun foods, as you would a doughnut or other calorie-laden, sugary dessert. Remember, people who consume sugary beverages do not feel as full as if they had eaten the same number of calories from solid food. Studies also show that those who drink sugary beverages do not counterbalance their high-caloric content by eating less food.

I should add that you would also do well to avoid or limit artificially sweetened drinks. These beverages can be even more damaging than those made with real sugar as a result of the effects artificial sugars have on the microbiome, resulting in a dysfunctional metabolism and higher risk for insulin resistance.[144] Also aim to reduce your consumption of fruit juices, even if they are 100 percent "pure and natural." A morning glass of orange or other fruit juice is not a bad idea, but consider that a twelve-ounce glass of orange juice, organic or not, contains about nine teaspoons of sugar, nearly the same as a twelve-ounce can of regular soda. If you insist on drinking fruit juice, choose unsweetened juices that contain no added sugars (or fruit juice concentrate, which is a euphemism for added sugar). Better yet, go for a pulpy vegetable juice. When choosing between a sugary beverage and bottled water, go with the water.

RULE #5: CHOOSE A WEIGHT LOSS DIET THAT WORKS FOR YOU

The byline spoke volumes: "Eating fatty foods has a 'shocking' effect." Indeed, the health news media was quick to cover Dr. Matthew Rodeheffer's 2015 paper,[101] which completely changed how we think about the effects of sitting down to a high-fat meal (let alone many high-fat meals). Dr. Rodeheffer is a member of the Department of Molecular, Cellular and Developmental Biology at the Yale University School of Medicine. Rodeheffer and his colleagues showed that in contrast to what was previously believed, a high-fat diet increased the number of fat cells before the existing fat cells were filled. That is, the mechanisms controlling fat cell number were thwarted by overdoing the consumption of fat. Worse yet, fat consumption led to more fat cells where you definitely do not want them to be: in the visceral fat depot. We do not know yet whether these new fat cells remain after you revert to a low-fat diet, but our work with TBT suggests that once the number of fat cells has increased, the animals remain fat even when diet is returned to normal for several months. Rodeheffer's results shifted how I view high-fat diets such as Atkins or Paleo and its variants. When you eat these high-fat diets, are you unwittingly triggering the fat machinery in your body to turn against you? This needs to be tested in carefully controlled clinical experiments. But until we have the definitive results, I think it is best to avoid high-fat diets. This is not simply a matter of weight. There are long-term consequences to triggering the body to create new visceral fat cells that may be there permanently.

Weight loss diets abound today, and I am frequently asked

which one is "best." To many people's surprise, and contrary to the trends constantly circulating in health circles, I do not believe there is one diet that fits everyone. If we accept the idea of personalized medicine, and that we are all different to some biological—and metabolic—degree, then each of us must find what works for best for us, so long as we base our dietary choices on unadulterated, fresh foods. As I outlined earlier, everyone responds uniquely to identical foods. Recall the 2015 Israeli study that tracked the blood sugar levels of eight hundred people over a week and found that the same exact diet in different people led to radically dissimilar effects on blood sugar.[19] The findings demonstrated that personalized medicine extends far beyond pharmaceuticals and encompasses diet, because how the same foods are metabolized changes from one person to another. This strongly supports the power of personalized nutrition in helping each of us to identify which foods can help or hinder our health and weight loss goals.

To be fair, it is definitely true that if you eat fewer calories than your body needs to sustain your level of activity, you will lose weight. The popular press is littered with fad diets that will help you lose weight: the ice cream diet, the Twinkie diet, the Hollywood cookie diet, the cabbage soup diet, the baby food diet, the grapefruit juice diet, and so on. These will all work to some degree, just as long as you consume fewer calories than you burn. However, as soon as you stop dieting and your caloric intake increases, you will gain weight and return to at least your programmed weight, if not higher. Remember that more than 83 percent of people who lose substantial amounts of weight will gain it back in a few years.[38,39] Consider the fate of *The Biggest Loser* contestants who worked so hard to lose weight but gained it back.[57] You can't diet to create sustained weight loss. As television personality Richard Simmons used to say, "You have to live it." What he meant was that the only way to change how your body looks in a permanent way is to permanently change the way you eat and exercise. Perhaps author and food activist Michael Pollan gave the most simple and easy-to-follow advice: Eat real (whole, unprocessed) food, not too much, mostly from plants.[197]

Toward this end, if you want to lose body fat, in addition to the previous tips about avoiding obesogens, I recommend experimenting with eating habits to find what works for you. While I am not a "diet doctor," I think that the following tips make good sense and have strong scientific support:

- Just as you would do well to avoid high-fat diets, skip low-fat diets. Skip any kind of "diet" that restricts one or another type of essential nutrient (fat, protein, or carbohydrate) claiming to promote sustained weight loss. Low-fat protocols often have people eating too many carbohydrates and refined sugars. There is value in having the right amount of healthy fats in your meals to offer flavor and promote satiety. Healthy fats include the monounsaturated fats found in extra-virgin olive oil, avocados, and certain nuts such as almonds and cashews; and the polyunsaturated omega-3 fats found in cold-water fish (such as salmon and sardines), grass-fed beef, chia seeds, flaxseed, pine nuts, walnuts, and Brazil nuts. Replace margarine and other butter-like imitators sold as vegetable oil spreads with organic butter or ghee (clarified butter that works well when cooking at high temperatures). Coconut oil has become popular with the Paleo crowd lately and can generally be considered a healthy fat despite its high saturated fat content. Coconut oil is rich in medium-chain fatty acids, which are easily digested and can increase HDL (good) cholesterol levels. The American Heart Association has recently published an opinion piece that classifies coconut oil as an unhealthy saturated fat.[198] Interestingly, the AHA article neglected to mention that the AHA receives funding from Bayer's Crop Science Division, the producer of LibertyLink soybeans, when they extolled the virtues of polyunsaturated oils such as soybean oil.[199] Journalist Gary Taubes wrote a blistering response to the AHA opinion piece.[200]

- Try intermittent fasting, which is defined as planned pauses in eating that can be anywhere from fourteen to twenty-four hours, once or twice a week. Much of this fasting time can occur during sleep (for instance, not eating from seven p.m. one evening until breakfast or lunch the next day). Although the prevailing medical wisdom has always been to eat frequent small meals, recent findings indicate that the opposite is best for weight control.[159,160] How intermittent fasting affects our metabolism is not entirely clear, and more research is necessary. We do know from animal studies that intermittent fasting may have a positive effect on blood glucose levels and on the ability to metabolize fat, particularly bad visceral fat that increases risk of obesity and chronic disease. Moreover, as we saw in chapter 5, mice that ate exactly the same number of calories in three meals vs. whenever they wanted gained less fat on the three-meal plan.[159]
- Eat more of your daily calories before three p.m. and avoid gorging at dinner and in the evening hours. Try not to eat within three hours of bedtime. The power of eating lunch

before three p.m. was highlighted in 2013 by a consortium of researchers from Harvard, Tufts, and the University of Murcia in Spain. They conducted their study in the Spanish seaside town of Murcia, where Spaniards make lunch their main meal of the day. To their surprise, the researchers found that when total calories consumed daily, levels of activity, and sleep quantity were equal, those who ate lunch later in the day struggled more with weight loss.[201] All 420 participants in the study were either overweight or obese. Each was put on the same twenty-week weight loss program. Half of the participants ate lunch before three p.m., and the other half ate after three p.m. Over the course of the twenty weeks, the early lunchers lost an average of twenty-two pounds, while the later lunchers shed only seventeen and at a slower clip. The mechanism appears to be related to how effectively fat is mobilized from the adipose tissue.[202]

- Try to plan your meals and snacks in advance. Set aside a day, perhaps over the weekend, when you figure out your eating schedule for the upcoming week based on your calendar in terms of work and personal responsibilities. Then write out your grocery list and do as much shopping ahead of time as makes sense so that you avoid scrambling to find food to eat for breakfast or to pack for a lunch, which can lead to eating out.

During certain times of the year I travel a lot, so I know how challenging it can be to eat healthily while on the road. Fast food can easily beckon, I am often treated to elegant restaurant meals, and the stuff they feed us on airplanes bears only a passing resemblance to food. When traveling abroad, I try to eat the best fresh, local food using advice from colleagues or online tools to find excellent restaurants (I am a big fan of TripAdvisor). When traveling in the United States, I also try to find the best fresh, locally sourced food and to leave some on the plate, since we typically serve much larger portions here than abroad. Always be mindful of portions whether you are preparing foods yourself at home or eating out. Portion sizes have exploded during my lifetime. We have become accustomed to "supersizing" our meals and going back for more. We expect to get big portions—it is practically built into our culture now. If you have a well-balanced meal with a high-quality protein, complex carbohydrates, and healthy fats, you should have less trouble regulating your intake. Eating slowly and savoring your food often helps to prevent overeating.

Being an informed, smart consumer is not really difficult or challenging. The rules in this chapter will go a long way to get you started on a healthier path free of obesogenic and endocrine disrupting chemicals. Following these guidelines will almost certainly reduce the toxic load of hazardous chemicals in your body and should help you to see—and feel—positive results in your life, from your waistline to your outlook.

CHAPTER 7

Step 2: Purge the Plastic

Reducing the Plastic in Your Life

Plastic is everywhere. From cars to computers, bath toys to bottles, clothing to kitchen tools and storage containers, plastic is pervasive in our lives. Even though I implore people to avoid it, even I cannot live 100 percent plastic-free, a nearly impossible feat in the twenty-first century. In the last decade we produced more plastic than during the entire twentieth century. Fully half of the plastic is used just once and thrown away. Meanwhile, our bodies (as well as the environment) bear the brunt of effects from plastics, some of which are permanent.

Clearly, it is not feasible to live in a bubble, nor should we drive ourselves crazy trying to control everything by changing our lifestyle overnight. Balancing our legitimate concerns about chemicals in plastic that could be adversely affecting our health, including metabolism and weight control, with our dependence on the convenience of plastics in our daily lives (at least to some degree) is the subject of this chapter. Let's get to the bad news first and then we will move on to the empowering information that will help you do something other than just worry.

THE BANE OF THE BOTTLE

Plastic water bottles are among the most problematic conveniences in our lives today for a variety of reasons. Their manufacture to meet our demand in the United States alone uses more than seventeen million barrels of oil annually, not including the energy for transportation. The average American uses more than 160 plastic water bottles a year but recycles only about 38 of them.[194] As I was proofreading this chapter, a study was published revealing that an estimated thirty-eight million pieces of plastic trash were found on Henderson Island, an uninhabited, formerly pristine island in the middle of the South Pacific Ocean.[203] It is a great idea to keep plastic out of landfills, but it may use more energy to recycle plastic containers than to make new ones—it is better to avoid them altogether.

In comparison with what you might pay for drinking bottled water, the cost of having tap water delivered to our doors is very inexpensive. This is even true in Southern California, where the cost in San Diego is about $0.01 per gallon for most residential customers. The idea that bottled water is cleaner and comes from higher-quality sources (cue the image of snow-capped mountains and running streams) is not necessarily true. Many of the most popular bottled waters are purified municipal tap water; just check the bottle and see whether it contains "purified water" (a euphemism for tap water) or spring water bottled from a defined source. The primary purpose of such filtration is to remove the taste of chlorine and objectionable minerals such as iron, not to remove hazardous chemicals. Worse, no matter how pure the water started off, the longer it stays in plastic bottles, the higher the concentration of chemicals leached from the bottles it will contain, especially when it has been stored and/or transported warm. So what kind of obesogens hide in plastic?

EDCs IN PLASTICS

Among obesogenic chemicals in plastic that wreak havoc on the human body and concern me the most are endocrine disrupting chemicals (EDCs). As I detailed earlier, EDC exposure is especially worrisome when it occurs during fetal development or early life. These are "windows of susceptibility," when children are physically developing and their small body masses cause them to be more susceptible to smaller amounts of harmful substances than are adults. EDCs can easily enter the body through inhaling household dust, eating pesticide-filled foods from plastic storage containers, and using personal care products containing phthalates, parabens, and "fragrance." Unfortunately, only a bit more than two hundred chemicals are measured in NHANES, meaning that a large number of chemicals are not examined (including the majority of known EDCs). Therefore, the number of exposed individuals, as well as the typical levels of exposure, remains unknown. Relatively little sampling of infants and young children occurs in NHANES because most parents are understandably unwilling to subject their children even to blood and urine testing. Conducting such studies is also extraordinarily complicated as a result of the high cost and extensive bureaucracy involved when working with human subjects, particularly children.

Bisphenol A is among the most ubiquitous of EDCs. BPA is the starting material for making polycarbonate plastics and is widely used in a variety of plastic products ranging from bottles and food can linings to toys and water supply lines. BPA is released into the environment and routinely ingested when these plastics degrade. Containers made from or lined with BPA and its relatives can leach BPA and contaminate the contents. Proponents of BPA use often argue that it is rapidly broken down by the body, rendering it harmless, and also note that BPA is not stored in the body, unlike other chemicals.[204] This argument has several flaws that the informed consumer should be aware of.

While BPA does not get stored in the body, we are exposed almost constantly, making this a moot point. Although it is true that BPA in your food and water supply can be rapidly bound to a sugar group in the intestine and liver (BPA-glucuronide), BPA can also be directly absorbed into your blood through the skin in your mouth and esophagus (the oral mucosa), through your skin when you touch thermal paper receipts (which are also ubiquitous), and through your lungs when you breathe BPA dust from these same thermal papers. BPA that gets into your body this way bypasses metabolism in the liver and intestines and is available to cause trouble throughout the body. Scientists at Health Canada have shown recently that the so-called breakdown product of BPA, BPA-glucuronide, is a potential obesogen that causes cultured cells to differentiate into fat cells.[205] Think about these issues the next time you hear an industry spokesman or apologist tell you how BPA is harmless and rapidly eliminated from the body.

There has been a consumer-driven push to remove BPA from products, particularly those to which children are exposed, such as sippy cups and baby bottles. This is a shining example of how powerful the voices of concerned moms are. The BPA controversy was ignited by University of Missouri biology professor Frederick vom Saal, who together with his colleagues showed that exposure to low levels of BPA—in the range of what is found in people—can harm the prostate.[206] Fred and his colleagues followed this with a series of studies showing effects of BPA at low doses in a variety of adverse outcomes.[207-211] Since then, a multitude of other studies have created a large body of literature that links low-dose BPA exposure—exposures much lower than what had been previously deemed "safe" by the FDA—to much more than just prostate damage. Not until 2007 did the government begin to take action, starting with an investigation into its own wrongdoings: Sciences International, the firm hired to review BPA toxicity for the government, was found to have corporate clients such as Dow Chemical and BASF that are major manufacturers of BPA.[212] To say there was a conflict of interest is an understatement. And not until pressure from Congress in the spring of 2008 (which was provided by Health Canada declaring that BPA was toxic and should be removed from baby bottles) did the FDA admit that BPA might pose a risk to humans. That was when retailers such as Walmart started to pull BPA products from their shelves and we began to see a new market for "BPA-free" products.

Unfortunately, despite this success, BPA remains ubiquitous in our society, and its "new" replacements—such as bisphenol S (BPS) and bisphenol F (BPF)—appear to be just as hazardous based on a systematic review of the literature by scientists at the NIEHS.[213] BPS and BPF are examples of what I call the industry "whack a mole" game—when we ban or pressure industry to eliminate one chemical, they simply substitute a closely related chemical for which there are fewer data available (in other words, you whack one mole and another one pops up, just like in the game). These alternatives to BPA were assumed to be more resistant to leaching or somehow less toxic than the original chemical, and therein lies the problem: ***assumed***. Unfortunately, many chemical replacements are not extensively tested before being placed on the market and are similar enough to the original chemical that one might reasonably expect they would cause about the same effects. Such is the case with BPS and BPF. Sadly, the legal burden in the United States is on the EPA and other government agencies to demonstrate that the chemicals are hazardous, rather than on the company to show they are safe.

My colleague Professor Laura Vandenberg of the University of Massachusetts–Amherst is one of the "young guns" in the EDC field. Laura is among the trailblazing scientists who are trying to end the controversies surrounding BPA and its analogs. Laura did her graduate work at Tufts University with another friend and colleague, Professor Ana Soto—one of the founders of the EDC research field. Through their academic publications and subsequent press in the media, Ana and Laura sounded the alarm about BPA and breast cancer, stressing that it was wise to follow the "precautionary principle," especially when considering vulnerable populations including women, their fetuses, and developing children. The precautionary principle holds that when the effects of a chemical are not known or are disputed, it is better to avoid exposure than to suffer the consequences later. In her own lab at UMass, Laura's research continues to incriminate BPA and its analogs, and more recent research has shown that low-dose exposure to BPS, particularly during pregnancy and lactation, affects maternal behavior in mice as well as how the brain is "wired."[214] This stunning finding was highlighted in a 2017 review aptly titled "The 'Plastic' Mother."[215]

BPA and BPS are not the only bisphenols out there. Shifting from BPA to BPS to BP-whatever is not the solution. And the bisphenols are not the only EDCs found in plastic. We don't even know all of the potential EDCs out there today in plastic. In 2013, for instance, German researchers led by Dr. Martin

Wagner at Goethe University Frankfurt used an unbiased, functional test to identify a new EDC in bottled water sold in polyethylene terephthalate (PET or PETE) containers.[216] Martin and his colleagues tested extracts of water from various manufacturers in a series of assays that measured the estrogenicity or androgenicity of the chemical and discovered several chemicals not even known to be in the bottles. Granted, there have not yet been detailed studies about how these chemicals affect human health, but we know quite a lot about the effects of estrogens and antiandrogens in the body. Studies such as these highlight the need for more research amid the astonishing gap in knowledge of EDCs and other chemicals that leach from food contact materials. It's worth noting, however, that Martin and his colleagues did not find these EDCs in water samples from some of the same manufacturers that were instead packaged in glass bottles.

The best we can do as citizens is to demand proper testing before approval of chemicals (a worthwhile but perhaps unattainable goal). It behooves us to avoid any suspicious materials as much as we can in our own lives—to practice a "personal precautionary principle." It can take a long time to rid the environment of these chemicals, even once they are banned (which rarely happens). For example, people continue to test positive for polychlorinated biphenyls (PCBs), a class of EDCs banned in 1979. Yet given the hazards of exposure, it is certainly a fight worth having.

WHAT'S IN A NUMBER? DECODING PLASTIC

While I do not expect you to memorize the classification system used by the plastics industry to categorize their wares based on chemical makeup, it helps to know what you are looking

Resin	Resin Indentification Code-Option A	Resin Indentification Code-Option B
Poly(ethylene terephthalate)	1 PETE	01 PET
High density polyethylene	2 HDPE	02 PE-HE
Poly(vinyl chloride)	3 V	03 PVC
Low density polyethylene	4 LDPE	04 PE-LD
Polypropylene	5 PP	05 PP
Polystyrene	6 PS	06 PS
Other resins	7 OTHER	07 O

at—for the most part—when you see those numbers. All modern plastics are composite materials: they are mixtures of ingredients rather than a single chemical. However, there are broad groups into which plastics can be categorized. Here is your cheat sheet:

Code 1: Made with polyethylene terephthalate, also known as PETE or PET. Items made from this plastic are used once and then recycled into new, secondary products such as carpet. PET plastics are commonly used for beverage bottles, peanut butter jars, medicine jars, combs, rope, and beanbags. PET-based containers sometimes absorb odors and flavors from foods and drinks that are stored inside of them.

Code 2: Made with high-density polyethylene, or HDPE. Items made from this plastic include containers for milk, shampoos and conditioners, soap bottles, detergents, bleaches, and motor oil. Many toys contain this type of plastic as well. Recycled HDPE is used to make plastic lumber and crates, fencing, and more.

Code 3: Made with polyvinyl chloride, or PVC, which is not often recycled and is used for plumbing pipes and floor coverings. PVC can also be found in bibs, mattress covers, and a few types of food and detergent containers. PVC can be harmful if ingested and should not come in contact with food items. PVC may also contain phthalates to soften it.

Code 4: Made with low-density polyethylene, or LDPE, which is not typically recycled. LDPE is durable and flexible; therefore, it is often found in food storage items such as plastic cling wrap, sandwich bags, squeezable bottles, and plastic grocery bags. Recycled LDPE is used to make garbage cans, lumber, furniture, and many other household products.

Code 5: Made with polypropylene, or PP, which is not recycled as much as the PET- and HDPE-based plastics. PP is strong and able to withstand high temperatures; therefore, it is used to make containers for food storage, margarine, and ice cream and yogurt. PP is also common in plastic bottle caps, syrup bottles, prescription bottles, and some stadium cups. Recycled PP is used to make durable tools such as ice scrapers, rakes, and battery cables.

Code 6: Made with polystyrene, also known as PS and Styrofoam, which is not recycled efficiently. Examples: disposable coffee cups, plastic food boxes, plastic utensils, and packing foam. Recycled PS is used to make many different kinds of products, including insulation, license plate frames, and rulers.

Code 7: This category is where all the miscellaneous types of plastic are placed. It is the grab bag and includes polycarbonate (PC), which is a plastic made from BPA (new plastic alternatives to polycarbonate are also marked Code 7). Many plastics in this category are difficult to recycle. Examples of PC: baby bottles, multigallon water bottles, clear plastic cutlery, sports bottles, compact discs, and medical storage containers. Plastics that can be recycled from this category are used to make plastic lumber, among other products.

The "safest" plastics in my view are the ones coded 1, 2, 4, and 5. Avoid those coded 3, 6, and 7 unless you are certain that the Code 7 plastic is one of the new, compostable green plastics made from corn, potatoes, tapioca, or rice. Unfortunately, nontoxic, biodegradable plant-derived plastics currently get the same Code 7 as BPA-based plastics, which can easily confuse consumers. The best approach is to avoid plastic as much as possible.

HOW TO REMODEL YOUR PLASTICIZED LIFE

In my home, we use glass, stainless steel, or porcelain for nearly anything that comes into contact with food or heat or can be ingested. Food storage can be tricky because plastic food storage containers, plastic wrap, and plastic bags, including freezer bags are so convenient. While we mostly avoid plastic, I confess to using plastic wrap to cover some containers and plastic freezer bags for dry items (though I try to freeze liquidy foods in glass bowls). Plastic lids are sometimes unavoidable, too. Note that rubbery silicone may not be a safer alternative. Silicone is used in a lot of kitchenware today, from oven mitts, tongs, and pan handles to nonstick baking sheets and parchment paper. Some tips:

- This one bears repeating: Give up bottled water and other beverages in plastic containers for good. Get a glass or stainless-steel travel mug or bottle. I use a stainless-steel sports bottle that keeps my water cold.
- Use common sense when deciding how to store your foods and beverages. Use glass, ceramic, or stainless steel whenever possible, especially for liquids. Foods frozen in plastic containers will leach fewer contaminants than those stored at room temperature or in the refrigerator (or heated in plastic). The ability of chemicals to leach out of plastics is temperature dependent, for the most part.
- Never, ever, microwave, cook, or bake using any plastic.
- When buying nonstick cookware, look for ceramic-based coatings, but be sure to purchase from reputable manufacturers who will vouch for the safety and authenticity of their products. In the United States, the FDA requires manufacturers to certify that their ceramic products are free of lead or cadmium. Unfortunately, this is on the honor system—they do not police it. There have been reports that some ceramic glazes leach these chemicals. Beware of cheap imports, as the manufacturers often cut corners. Never heat cookware with nonstick coatings on high heat, as you risk decomposing the coating into toxic

vapors that you will breathe (this is particularly bad for various forms of PTFE, aka Teflon coatings).

- When purchasing groceries, try to avoid canned goods and items sold in plastic (frozen produce in plastic gets a free pass). Buy cereal, pasta, rice, nuts, and seeds from bulk bins and fill a reusable bag or container.
- Avoid water filtration systems that employ plastic pitchers. As noted in chapter 6, it is best to install the filter at the source of your water.
- When buying toys for your children, choose wood with nontoxic paint rather than any type of plastic, even items labeled "BPA-free." Avoid any toys, pacifiers, and teethers that list PVC as ingredients. Watch out for "hospital-grade silicone," too. While pacifiers labeled "BPA-free" may employ silicone instead, the replacement may not be any better. Silicone contains organotins, including tributyltin—the obesogen that got me into this research to begin with. Choose bottles made of tempered glass, polypropylene plastic, or polyethylene plastic.
- Avoid those microplastics in many personal care products such as exfoliating facial scrubs by not buying items with "polypropylene" or "polyethylene" on the ingredients list.
- Avoid phthalates, which are a class of compounds known to be obesogens for their endocrine disrupting effects. Phthalates are used in the making of a wide variety of soft, flexible plastic products including vinyl (PVC) goods (they are also added to lots of personal care products to make their fragrances last longer; see the next chapter). Look for goods marked "phthalate-free," and if you find

yourself in a medical setting that involves medical tubing, catheters, and blood bags, ask for phthalate-free tubing.
- Bring your own thermos or mug to the coffee shop when you can.
- Bring your own reusable shopping bags to the grocery store (in some states, such as mine, plastic bags have been banned). When you are empty-handed, choose cardboard boxes or paper bags over plastic bags if you have a choice.

The benefits of reducing the amount of plastic we use go far beyond avoiding their harmful, weight-inducing ingredients and include preserving the environment. Plastic does not biodegrade like other organic materials such as plants and paper. Instead, it breaks into smaller and smaller pieces over time, never really going away. Because of this, nearly every single piece of plastic ever produced is still present today somewhere. Every year, more than three hundred million tons of plastic are made around the world, and only 10 percent of all plastic used is properly recycled. The threat to the environment from this vast accumulation of plastic waste is increasing daily. The gyres at the centers of each major ocean on earth are giant garbage patches filled with plastic (recall the earlier example of Henderson Island).[203] Birds, fish, and other wildlife can mistake plastic bits for food, become entangled in plastic debris, and so forth. More than one hundred thousand marine animals and one million birds die every year from ingesting and choking on plastic. Pests such as the mosquitoes that carry Zika virus and West Nile virus thrive in water found in discarded plastic containers, unlike more common types of mosquitoes that breed in ponds, streams, and lakes.

Production of plastic products emits millions of tons of

greenhouse gases that contribute to global climate change (even the process of recycling plastic is highly suspect, as it can also be polluting). Chemical breakdown products of discarded plastic can leach into the soil and end up in our food supply. By eliminating the unnecessary plastic from your life, not only will you be helping your waistline, you will be helping the planet.

CHAPTER 8

Step 3: Rethink Your Personal Space

Your Simple Household Makeover

In chapters 6 and 7, we discussed ways to reduce your obesogen and EDC exposure that are relatively simple and give you large reductions in exposure for relatively modest efforts on your part. Avoiding obesogenic ingredients in your diet and minimizing your use of plastic will go a long way to protect you. In fact, if you focus on just those two goals alone for a while, you can pat yourself on the back. Then, when you are ready to take your cleaner living another step forward, consider the advice in this chapter.

When I was growing up, seat belts (if you could find them) were optional, the drinking age was eighteen, marijuana was illegal everywhere, and people could smoke wherever they wanted, airplanes included. Today these behaviors are either banned below a certain age or totally prohibited. There were lots of other things we did as kids that would be frowned upon today or otherwise seen as unhealthy. Everyone microwaved foods in plastic, ate trans-fat-filled margarine, rode bikes and motorcycles with no helmet, and drank water from the phthalate- and lead-leaching hose in the backyard.

Each generation finds new dangers to regulate, and I expect we will see more chemicals and their associated products undergo scrutiny and testing. Sadly, regulation lags far behind such investigations. This has two main causes. First, the standard that federal regulators must reach to remove a chemical from the marketplace requires that the chemical be demonstrated "to a substantial certainty" to cause harm to humans. This is a virtually impossible standard to meet, because as mentioned earlier, it is illegal, immoral, and unethical to do the types of controlled exposure studies on humans needed to definitively prove that any chemical caused an adverse effect in a human. I have never met a lawyer who could explain to me why the chemical law uses the substantial certainty standard, why chemicals have more protection than people. In civil law, the standard is "by a preponderance of the evidence," and in criminal law the standard is "beyond a reasonable doubt." Perhaps you would not be surprised to learn that the chemical industry and industry trade organizations have spent millions, if not billions, of dollars creating doubt about the safety of their products to counter the types of careful science done by independent, publicly supported scientists. For a good discussion of this topic, you would be greatly enlightened by reading *Merchants of Doubt* by Naomi Oreskes and Erik Conway.[217] As a result of industry doubt-mongering (which you can observe in the recent discussion about the safety of the herbicide glyphosate), it is not wise to depend on government regulations to protect our health and safety. Even if there were adequate regulations, someone would need to be enforcing them, and one wonders who would do this. The EPA has dropped the ball entirely.

Obesogens lurk in some of the most commonly used everyday products, from cleaning solutions and air fresheners to body lotions, cosmetics, and bubble bath. By the time we find out about the potential hazards of a substance (or behavior or activity), many of us have already experienced exposures and their effects. The EPA, European Union, and World Health Organization have promised to speed up their efforts to gather data on "contaminants of emerging concern," but as already noted, it is very unlikely that any regulations will be implemented fast enough to ensure our safety. It behooves each one of us to shop around, avoiding products that are sources of potential toxins to the body, especially EDCs and obesogens.

YOUR BODY BURDEN

One thing you are likely wondering about now is how many potentially harmful chemicals are in your body. Although scientists have been measuring industrial pollutants for decades in our environment, only recently have we begun the process of monitoring the so-called body burden, the levels of synthetic chemical toxicants in tissues of the human body. This biomonitoring, for which blood, urine, umbilical cord blood, and breast milk are analyzed, is being conducted by several high-profile institutions and research organizations on an ongoing basis. The Centers for Disease Control and Prevention (CDC) runs the National Health and Nutrition Examination Survey (NHANES), which gives us a snapshot of the body burden of about two hundred toxic chemicals in a small sample of the U.S. population (typically a few thousand people statistically selected to represent the approximately 325 million total population). In the last decade, the new Division of Environmental Hazards and Health Effects in the CDC has established a national system for tracking environmental hazards and the ailments or illnesses they may cause. The National Institute of Environmental Health Sciences, established by the NIH in 1966, also undertakes and supports research, but it is not involved in the biomonitoring.

Bear in mind that body burden studies such as those by the NHANES report averages according to sex, age, and race in a very small number of people. It should be obvious that this neither measures nor predicts body burdens in other individuals and does not consider all possible contaminants or the potential for synergy among chemicals. Everyone reacts differently to external stimuli, including the effects of combinations of various stimuli such as chemicals. As a result, regulatory standards for limiting exposures to known pollutants in food and water will probably not protect uniquely vulnerable populations, such as children or people of any age with chronic illnesses or subsets of the population that may be more sensitive than average. There is also a nearly complete lack of regulation in the manufacturing industry that creates new materials to be used in products. As a reminder, no federal agency tests the toxicity of new materials before they are allowed on the market. Instead they rely on the manufacturers to perform basic toxicity testing, which is a fundamental and unacceptable conflict of interest.

Note: "Toxins" are poisons derived from biological sources (that is, poisonous substances produced by living cells or organisms). In contrast, "toxicants" are man-made, even if they resemble toxins (for instance, pyrethrins, which are natural pesticides produced by chrysanthemums, compared with pyrethroids, which are synthetic compounds related to pyrethrins). As we already discussed in chapter 3, toxicologists (scientists who study the effects of poisons) have a different view about what a toxicant is from that of the rest of the scientific community. Most of us can readily distinguish poisons from essential substances, but to many toxicologists, every substance in the world is a poison of some sort. What a depressing (and inaccurate) worldview!

People living in industrialized nations now have hundreds of synthetic chemicals in their bodies accumulated from food, water, and air.[218,219] The vast majority of these chemicals, many of which come from plastic, have never been adequately tested for health effects. Chemicals from plastics can be absorbed by the human body—93 percent of Americans six years of age or older test positive for bisphenol A, which you know now is a chemical derived from plastics and a known estrogen and obesogen. Some of the other compounds found in plastic have also been found to affect hormones or have other potential human health effects. Among the most alarming studies done on the body burden are those that show how vulnerable pregnant and lactating women can be to chemical exposures. My late friend and colleague Professor Howard Bern of the University of California–Berkeley, one of the giants in the field of endocrinology and reproductive biology, first wrote about the concept of the "fragile fetus" in 1992,[220] a must-read for anyone interested in this field.

Although we once thought that the placenta acts like a shield, protecting the fetus from most chemicals and pollutants in the environment, we now know differently. Industrial chemicals and pollutants such as residues from cigarettes and alcohol can indeed stream through the placenta. Benchmark studies spearheaded by the nonprofit Environmental Working Group first demonstrated this when they found 287 different chemicals in umbilical cord blood and breast milk. Incredibly, these include 180 carcinogens, 217 central nervous system toxicants, and 208 known to cause birth defects in animals.[218,219] Even the blood–brain barrier can be penetrated in utero.

Think about this for a moment: none of these women or their children gave informed consent for being exposed to these chemicals. Those exposures could not occur without informed consent if they represented pharmaceutical drugs or medical procedures. Worse yet, there is no readily available, universal test you can use to determine what your personal body burden is—how many chemicals your body harbors and which kind. While you can probably order blood tests to search for certain chemicals, this is expensive, incomplete, and largely impractical unless you have a good idea which chemicals you have been exposed to. Other, larger studies have reported similar results—contamination of moms, fetuses, and children with toxic chemicals is widespread.[221,222] Is this a situation we should accept? How should we deal with this? I think that a major focus should be on preventing more exposures.

A particularly innovative approach to inferring blood levels of various chemicals in small children was developed by Professor Åke Bergman from Stockholm University and Swetox. Åke is an "out of the box" thinker, so he and his PhD student Jessica Norrgran reasoned that while people might be reluctant to have scientists sample their children's blood, they would have no qualms about having blood samples taken from their pet house cats, which spend much of their time on the floor and in a very similar environment to that of infants and young children.[223]

Indeed, Jessica and Åke found that there was a close association between levels of persistent organic pollutants they measured in house dust and those in the blood of house cats (and presumably of children in the same homes).

WHEN IN DOUBT, TAKE IT OUT

Your personal body burden is unknown, and there are no practical, economical tests you can take to determine what your current burden is. But what you can do is lessen your exposures and, therefore, your body burdens. In addition to thinking about how and what you eat, think about what you clean your home with, what you spray in the garden, how you furnish and decorate your home, and what consumer products you bring home, including personal care products. Note that natural will not always be the solution, and synthetic ingredients are not necessarily bad. Plenty of all-natural ingredients can do damage. Likewise, just because something is synthetic (for instance, synthetic glycerin and stearic acid in soaps, detergents, and cosmetics) does not mean it is harmful.

The cosmetic and beauty industry is horribly underregulated, though this could change in the future. Cosmetic companies police themselves and have come under fire lately for misleading customers about their ingredients and their potential hazards. Many cosmetics and personal care products are made with phthalates and parabens and contain a wide array of other undesirable chemicals. In the United States, the FDA oversees cosmetics, but unlike drugs, cosmetics do not require extensive testing and approval before they can be sold.

Now that your diet and kitchen are cleaned up (I hope), let's see what you can do in other rooms of the house. I introduced the precautionary principle in the last chapter. It applies here as well: ***When in doubt, take it out.*** Following the tips below will be good not only for you and your family, but also for the environment and community in the broader sense. When you choose to buy one item over another, you increase demand on one side and lower it on the other. This affects the economics on a larger scale and induces change in the industry and marketplace. In this way, you can encourage changes in the practices of industry, agriculture, and manufacturing included. Also, remember that you have a voice. You can advocate for change in your local community by speaking up. The parents in Non Toxic Irvine have started a nationwide, grassroots movement to get toxic chemicals out of our schools, parks, and public spaces. Calm logic and reason, rather than hysteria, hyperbole, and scaremongering, have persuaded city leaders in Irvine, San Juan Capistrano, and Burbank, California; Naperville, Illinois; and elsewhere in the world to make legislative changes that provide better choices for their communities. Never underestimate the power of a group of passionate parents, but steer clear of what I call "the lunatic fringe"—the people who blame one or another chemical for every ailment known to the modern world. They may mean well, but ultimately, shrill voices, unpleasant tactics, and shouting about fringe science deprive them of the credibility they so desperately crave and cause legislators and regulatory agencies to ignore them as well as other more moderate and reasonable groups with the same goals. Meaning well is not equivalent to doing well.

The rest of this chapter is organized in a common sense fashion, beginning with some basic recommendations that can make a big difference and then progressing to more specific details regarding household good and toiletries. For more comprehensive lists about specific chemical ingredients and to get brand recommendations, visit the Environmental Working Group website at www.ewg.org.

TAKE SHOES OFF WHEN ENTERING YOUR HOUSE

In many countries, it is customary to remove your shoes upon entering a house. It shows respect for the home and its occupants. However, in many Western countries, including the United States, it is uncommon to leave your shoes at the door (or outside). But leaving shoes outside can be one of the easiest things you can do to avoid exposures to harmful substances, ranging from pathogenic bacteria, viruses, and fecal matter to toxic chemicals, including a whole panoply of obesogens, pesticides, and oil or petroleum by-products. Think about what you trek through, albeit unwittingly, while you are outside: you trudge through bird droppings, asphalt sealed with coal tar, dog waste, gasoline and oil residues, pesticides, and herbicides, to name a few things. Your shoes may be even more toxic than your toilet. Your shoes carry contaminated dust from nearby construction sites as well as chemicals recently sprayed on lawns, around the perimeters of houses, near public parks, and even on the sidewalks outside your home, particularly if you live in a board-managed community.

If you have young children who crawl or walk around on the floor all day, it is even more important to not wear your outside shoes inside the home. Children aged two years and under put their hands in their mouths an average of eighty times an hour, not to mention objects they find on the floor. Babies are prone to licking floors and routinely putting their fingers into their mouths; older kids play with toys on floors and find those lost crumbs and fallen foods all the more delicious. Consider getting an inexpensive shoe rack, or keep a large basket by the front door to house your dirty shoes, or carry them to a nearby closet designated for this endeavor. Better still would be to leave the shoes in a foyer or "mudroom" or even outside if they are protected from the weather. Have a clean pair of slippers or socks to put on after you remove your shoes if you prefer not to go barefoot. Ask guests to remove their shoes, too.

WASH YOUR HANDS FREQUENTLY

Once you have made a habit of taking your shoes off upon entering your home, make washing your hands the second thing you do when you get home. You should be washing your hands frequently throughout the day—not just after using the bathroom. Handwashing is one of the most reliable steps we can take to limit chemical exposures and to prevent the spread of germs and toxins. Estimates vary, but on average we touch our faces multiple times an hour and hundreds (some say thousands) of times per day without ever realizing it. Every touch transfers substances on the hands into our bodies through our mouths, eyes, and nose. To be sure, not every substance will make us sick, be obesogenic or otherwise harmful to our health, but plenty of nasty substances can find their way to us by hitching a ride on our hands.

Handwashing is even more important when preparing foods. Effective handwashing does not require any fancy or costly products, nor do you need to scrub up as you see doctors do on television. Plain soap and water is all you need, and be sure to rub thoroughly for at least twenty seconds. The emerging consensus is that washing with plain soap is better than using soap with added antibacterials and antiseptic agents. Some of these are problematic (for example, triclosan), and continued use can simply lead to resistant bacteria. Your grandma had it right—a simple soap is just fine.

USE A VACUUM WITH A HEPA FILTER

Some of the following ideas require more effort and money (as in buying new household items to replace problematic ones currently in your home). Until you can consider big-ticket purchases such as new sofas, flooring, and mattresses, get yourself a good vacuum with a HEPA filter if you don't already have one. HEPA stands for "high-efficiency particulate air." To qualify as a HEPA filter, the product must remove 99.97 percent of airborne particles measuring 0.3 microns or greater in diameter passing through them. To put 0.3 microns in perspective, by comparison a typical human hair ranges from 17 to 181 microns in diameter. A HEPA filter is trapping particles several hundred times thinner, including most dust, bacteria, and mold spores. Volatile organic compounds (VOCs) often adhere to dust, so a HEPA vacuum will help you minimize flame retardants, organotins, phthalates, and other VOCs in your household dust and you will avoid inhaling those toxic chemicals.

Some sources of VOCs in your household:

- Building materials (walls, floors, carpets, vinyl blinds, and furnishings)
- Paints, particularly those with antifungal additives
- Aerosol sprays

- Cleaners and disinfectants
- Air fresheners
- Hobby supplies (such as glues, adhesives, and permanent markers)
- Dry-cleaned clothing
- Office equipment (for instance, copiers and printers)
- Pesticides

When buying these products, look for low VOC or no VOC; they will be labeled and marketed as such.

People in developed nations often spend more than 90 percent of their time in indoor environments. These environments can be more toxic than the outdoors in many ways. Numerous studies over the past decade, including one meta-analysis published in 2016 by a consortium of U.S. institutions, have definitely shown that household air can be a toxic cocktail—often filled with dust that contains chemicals known to be toxic to the immune, respiratory, and reproductive systems, VOCs such as formaldehyde, low oxygen and high carbon dioxide levels, and combustion by-products such as soot and carbon monoxide.[224] Collectively, these are associated with cognitive and behavioral impairment in children, asthma, and chronic disease. The authors in the 2016 report write: "The indoor environment is a haven for chemicals associated with reproductive and developmental toxicity, endocrine disruption, cancer and other health effects." The culprits are largely PVC vinyl building materials (as in vinyl flooring and vinyl blinds), chemically treated home furnishings, wall coverings, personal care and cleaning products, synthetic fragrances, and electronic equipment that contain plastics. Many of these chemicals are also obesogens.[225,226]

If you have carpets, try to vacuum thoroughly and as often as you can (at least once a week). You can also consider adding HEPA air purifiers to rooms that you spend the most time in (living room, den, bedrooms, and so forth). There are lots of options now on the market to fit any budget, but I advise not to go for the cheapest filters. You can find quite a few reviews of product quality on the Web. Purchase from reputable companies that have been making HEPA filters for years. Some advertise they use ozone to purify the air, but that is not necessary and ozone can be toxic to your health (notwithstanding the fact that we ***need*** ozone twelve miles up in the atmosphere). Don't buy into that.

Other ways to reduce indoor air contaminants:

- Use exhaust fans wherever you have them in rooms such as the kitchen (while cooking), bathroom (while bathing, showering, or spraying personal care products), and laundry areas (while doing the laundry).
- Ban smoking indoors.
- Minimize use of candles and wood fires with synthetic logs. When purchasing candles, look for those made from organic beeswax, which are not petroleum based like most traditional candles. Use a snuffer to extinguish candles to avoid creating a sooty cloud when you blow them out. Better still, buy candles in glass jars that have lids and simply put the lid on when you want the candle to go out.
- When possible, naturally ventilate your home or apartment using open windows. Obviously if you live in a big city, there is a balance between indoor and outdoor

pollution that you need to consider. City dwellers might want to keep the windows closed during the day when pollution from vehicle exhaust is high and open them at night to reduce indoor pollution.

- Ban air "fresheners," spray, plug-in, or wick, from your home. These are laced with phthalates and other chemicals.
- Wipe windowsills with a damp cloth and vacuum blinds regularly. Damp-mop tile and vinyl floors and vacuum or dust-mop wood floors regularly, weekly if possible.
- Keep any toxic materials that you feel are necessary, such as glues, paints, solvents, and cleaners in a shed or garage—away from your living quarters.
- To enhance the air quality when it is very dry, add a humidifier to each bedroom. This can help you sleep better as well as keep your skin, nose, throat, and lips moist, which in turn helps your body be a stronger barrier against dust and germs. Beware of antimicrobial additives in humidifiers; some of these are known to be toxic. Steam humidifiers do not require microbicides. I use a few drops of bleach per gallon of purified water in cool mist humidifiers.

HOW GREEN IS YOUR GARDEN?

For many homes, the lawn and garden are ground zero for the most toxic and obesogenic substances. The same holds true if you live in an apartment or condominium. Some tips:

- If you have Roundup or any other synthetic pesticides or herbicides in your garden shed or garage, it is time to get rid of them. Now. There are few more illogical things one can do than spraying herbicides that are probably human carcinogens for purely cosmetic purposes. Would you rather have weeds or cancer? There are also a variety of less toxic organic herbicides that will get rid of weeds without giving you or your dog cancer. Look for ingredients that do not sound like synthetic agrochemicals. Some of the organic herbicides use citric acid, clove oil, cinnamon oil, lemongrass oil, and vinegar.

- Adding a thick layer of mulch (such as ground-up cedar bark) on your flower beds and gardens reduces weeds and decreases water use (it also improves the appearance, in my opinion). When mulching to smother or inhibit weeds, do not use plastic or treated products. Instead, cover the soil with organic matter such as compost, bark, natural wood chips, grass clippings, or straw.
- While it may seem quick and easy to spray toxic chemicals to kills pest insects, or to use a toxic chemical to kill garden snails, these are bad for you and for the environment. They have unintended consequences on birds and beneficial insects. There are many organic alternatives, such as salt, vinegar, and even beer. Don't forget the use of beneficial insects you can release into the garden that feast on aphids, mites, caterpillars, and other plant-consuming bugs and are harmless to people, plants, and pets.
- Avoid plastic pots and watering cans; choose pottery that is either unglazed or uses lead-free glazes and colors. Remember grandma's old metal watering can? Get one like it.
- Use organic soils, compost, and fertilizers, especially for areas where you are growing plants that you will eat, such

as vegetables, fruits, and herbs. If you have the space and time, start your own compost pile to add to your soil (you will find lots of resources online to help you start the perfect compost bins).

- Replace your traditional garden hose (that leaches lead, BPA, and phthalate) with an NSF (National Sanitation Foundation) certified, phthalate-free potable water hose. If you can get PVC-free, so much the better.

Organic gardening is easier than you think and is not costly; many of the tools and resources you need are likely already in your own kitchen. Once you have replaced your soils, fertilizers, and pesticides with wholly organic varieties, you will be well on your way to a safer, obesogen- and EDC-free garden. Have a look at the Sustainable Food Trust website for more information and ideas about how you can garden organically and sustainably.[183]

TOILETRIES, MAKEUP, AND PERSONAL CARE PRODUCTS

When it comes to toiletries, deodorants, soaps, and beauty products—most of which you put on your skin—aim to switch brands when it is time to buy again. While our skin serves as a barrier of sorts to the environment, its primary function is to prevent water loss and help control body temperature. Many chemicals can be absorbed through the skin, going directly into your circulation, bypassing first-pass metabolism in the intestines and reaching much of the body before being metabolized in the liver. Chemicals can also enter the body through other routes. Chemicals in eye makeup, for example, can get into the body through mucous membranes in the eye. Volatile chemicals (fumes) from hair spray, hair dyes, powders, and perfumes can be inhaled into the lungs, where they quickly get into the bloodstream. Lotions, sunscreen included, are absorbed through the skin. Lipstick is easily swallowed, and shampoo can gain entry through the skin and eyes.

Look for the genuine USDA Organic Seal and choose products that are safer alternatives (go to the Environmental Working Group's Skin Deep database for more information and resources).[227] Beware of the word "natural" used on items. Currently, this word is unregulated to the point of being meaningless—it does not guarantee the product contains only natural or organic ingredients. And it bears repeating: Choose products sold in glass rather than plastic whenever you can.

Your first step should be to avoid the following ingredients, many of which are potential EDCs or are otherwise unhealthy:

Triclosan and triclocarban: In 2016, the FDA banned these antimicrobial chemicals from hand and body soaps because they appear to have the potential to disrupt the endocrine system, particularly thyroid hormone signaling, and they were subsequently associated with adverse health effects in humans.[228] To my knowledge, this is one of the first examples of a chemical being banned from personal care products for being an EDC. While triclosan and triclocarban have vanished from commercial antibacterial hand soaps, they are still found in a wide variety of personal care and household products such as toothpastes, mouthwashes, facial cleansers, dishwasher soaps, deodorants, and cosmetics. Read the ingredients. Avoid cosmetics and body care products labeled "antibacterial"—remember that your skin has a microbiome that is better not disrupted. Instead of using antibacterial soap, wash your hands with normal soap and warm water. If you use skin disinfectant when it is not feasible to wash, select an unscented alcohol-based product.

Phthalates: Again, this class of chemicals is often added to personal care products to carry fragrances and helps lubricate other substances in the ingredients. It is found in perfumes, hair gels, shampoos, soaps, hair sprays, body lotions, sunscreens, deodorants, and nail polish. Look for the word "phthalate" in the ingredients list (though it can be hidden within the "fragrance" label). Beauty companies are starting to voluntarily remove these compounds from their ingredients and will say so in their marketing.

Parabens: As defined in chapter 1, these chemicals are obesogens due to their effects on the endocrine system. They are used as preservatives in many personal care products. Look for the words "methylparaben," "ethylparaben," propylparaben," "isopropylparaben," "butylparaben," and "isobutylparaben" in the list of ingredients. Choose paraben-free products. Often, a company that removes phthalates also removes parabens and will say so on the label ("phthalate- and paraben-free"). During the summer months, be mindful of the types of sunscreens you are using on yourself and/or your children. The EWG keeps a guide to sunscreens that will help you make the best decision. This can be tricky terrain, because "organic" sunscreens can lead to nasty burns if they do not do their job. Just because a sunscreen contains chemicals does not make it a bad sunscreen; but you do not need the parabens and phthalates to create a sunscreen that effectively blocks or filters UV radiation.

A Note About Sodium Lauryl Sulfate:

Sodium lauryl sulfate (SLS), also known as sodium dodecyl sulfate (SDS), and its siblings, sodium laureth sulfate (SLES) and ammonium lauryl sulfate (ALS): There has been a lot of debate in the blogosphere about these chemicals, and it has been claimed that they can be contaminated with cancer-causing solvents such as ethylene oxide and 1,4-dioxane.[xii] Sodium lauryl sulfate is a detergent, surfactant, and emulsifier used in thousands of cosmetic products, as well as in industrial cleaners. It is present in nearly all shampoos, hair color and bleaching agents, body washes and cleansers, makeup foundations, liquid hand soaps, and laundry detergents. While technically not an obesogen, it may have other health consequences and can be an irritant and cause skin rashes in some people. If you are sensitive, use caution when buying products with these chemicals and consider switching to cleaners and personal care products with natural soaps in them (made from organic oils) rather than synthetic detergents.

GENERAL HOUSEHOLD GOODS IN BEDROOMS, LIVING ROOMS, AND CLOSETS

I would not expect you to move out of your house after you finish reading this book and embark on a total remodel with

---

[xii] Depending on the manufacturing processes, lauryl sulfates may be contaminated with measurable amounts of ethylene oxide and 1,4-dioxane. The International Agency for Research on Cancer classifies ethylene oxide as a known human carcinogen and 1,4-dioxane as a possible human carcinogen. Ethylene oxide can also harm the nervous system, and the California Environmental Protection Agency has classified it as a developmental toxicant based on evidence that it may interfere with human development.

new furnishings, striving for a chemical-free home. Having said that, I know at least one writer who moved her entire family to the backwoods of Maine and is living off the grid in an attempt to avoid chemical exposure as much as possible. Here are some general tips to keep in mind for when you are able to implement them:

- When refurbishing your home, start with flooring. Carpets, even natural wool carpets, are magnets for dust and whatever toxic chemicals are adhered to them. Go for natural hardwood or ceramic tile for the lowest potential exposure. If you insist on carpets and have pets or small children, it is worthwhile to invest in natural fibers that have not been treated with stain-resistant chemicals. Synthetic carpets can off-gas chemicals for years that can affect the health of sensitive people, so select low VOC products. Hire an expert to remove the old carpet. The padding may contain polybrominated diphenyl ethers (PBDEs). If you are a do-it-yourselfer, protect your health by using gloves and an air-filtering mask. Vinyl flooring is less desirable than old-fashioned linoleum, which is made from natural ingredients including cork and linseed oil; importantly, linoleum does not contain PVC.
- The next time you can invest in a new couch or bed, choose goods made without toxic adhesives and glues (such as those containing formaldehyde), plastics, synthetic wood or particleboard, and treated wood. I realize that it can be difficult, if not impossible in many instances, to know exactly what is in your household goods and furniture because manufacturers will not necessarily disclose their presence (and your salesperson may not know or have

full access to such information). This is why it is important to buy from reputable brands that have long-standing track records. Look for a statement on the company website or sales literature that the product is certified to be low in VOCs. Ask for it in writing if you have any doubts. When you see a crazy bargain price that is too good to be true, proceed with caution. My daughter will be very angry when she reads the next statement because she is a devout vegan, but one way to reduce your exposure to chemicals found in furniture padding and stain-resistant treatments is to buy furniture covered with leather. This is largely impermeable to releasing materials from the padding, although you should be cautious about what types of dyes were used to color the leather—choose vegetable-based dyes whenever you can.

- Until you are financially prepared to buy a new mattress made from organic materials, the best you can do is purchase a barrier cover made of 100 percent organic cotton or wool. This will not block gases since it is a fabric, but it is better than using a plastic barrier that will itself off-gas. Your next best alternative is to look for a mattress cover made with a specially formulated polyethylene that has been tested and confirmed to block off-gassing and will not off-gas itself. And use hypoallergenic pillows filled with natural fibers such as cotton, wool, or feathers.
- When purchasing clothes, fabrics, upholstered furniture, or mattresses, choose items that are free of stain-resistant and water-resistant coatings (and avoid using sprays to make your purchased items stain-resistant and water-resistant/water-repellent). Many of these contain perfluorinated chemicals that are obesogenic and may be

hazardous to your health. Avoid reupholstering foam furniture. Try to avoid all foam if you can, though this can be tough. At a minimum, try for foam without flame retardants that are labeled as low or no VOC.

- Cleaners are often loaded with chemicals, many of which are linked not only to obesity, but also to cancer, allergies, asthma, and other respiratory ailments. When you buy household detergents, disinfectants, bleaches, stain removers, and so on, select green cleaning products with simple ingredients that have been around forever (for instance, white vinegar, borax, hydrogen peroxide, baking soda, soap). Be cautious when you see labels that say "safe," "nontoxic," "green," or "natural," because as I already mentioned, these terms have no legal meaning. Read labels carefully, identify the ingredients, and pay special attention to warnings. Obviously, you should try to avoid products labeled "poison," "danger," or "fatal" if swallowed or inhaled, particularly if you have young children in the house. Avoid anything with the following ingredients: diethylene glycol monomethyl ether, 2-butoxyethanol (EGBE), and methoxydiglycol (DEGME). For more details and product information, EWG is a good place to start.[229] You can accomplish a lot with harmless and highly economical ingredients to make your own cleaning products. There are thousands of easy recipes online that use old standard, nontoxic ingredients.
- Remember the "fragrance exception." According to federal law in the United States, the components of any substance labeled as "fragrance" do not need to be disclosed to the EPA, FDA, or any regulatory agency. If you want the inside story about this, rent the movie *Stink!*,

which documents Jon Whelan's quest to discover whether there may have been chemicals in household products that caused or contributed to the breast cancer that killed his wife at a very young age.[230] Note also that one of the industry-favoring characters in the movie, former U.S. congressman from California Cal Dooley, is now president and CEO of the American Chemistry Council—the most prominent chemical industry trade group that works tirelessly to prevent regulations that restrict chemical use.

- When looking for a dry cleaner, choose one that uses "green" technology. The main solvent used in conventional dry cleaning, perchloroethylene (or "perc"), has been deemed a hazardous air pollutant by the EPA and when disposed of must be treated as hazardous waste. Green, or non-perc-based, methods are better for you, for the employees of the dry-cleaning plant, and for the environment.

A Note About Flame Retardants:

Chemical flame retardants are common in consumer products. By law, they are added to a wide variety of household items such as furniture, electronics, appliances, mattresses, carpeting, clothing, and even baby products such as car seats, changing table pads, portable crib mattresses, nap mats, and nursing pillows. Flame retardants were required to be added to consumer products as a result of people burning themselves or their children in house fires caused by cigarette smoking. This was well-intentioned but ultimately deleterious to everyone, particularly firefighters, who breathe the toxic smoke these flame retardants produce while they battle fires. You can learn more about the industry-waged campaign to get flame retardants into many consumer goods in *Merchants of Doubt*, which I mentioned earlier.[217] The movie version has a particularly enlightening section about flame retardants. Many flame retardants are known or suspected obesogens, among their other deleterious effects. Worse, rather than remaining confined to the products that contain them, flame retardants migrate out of products and can contaminate house dust, which accumulates on the floor where children play and babies crawl. Flame retardants are nearly impossible to avoid completely, but there are simple precautions you can take to minimize exposure.

- Reminder: Use a vacuum cleaner fitted with a HEPA filter.
- Replace foam products made before 2005. Older foam items commonly contain PBDEs, highly toxic fire retardants that were taken off the U.S. market. Newer substitutes for those PBDEs, however, may not be any better, so try to buy products totally free of flame retardants.
- Choose clothes made from natural fabrics, and always wash new clothes before wearing them. When buying baby and children's clothing, be extra mindful because children are more sensitive to the effects of chemicals found in flame retardants. Look for clothes specifically labeled as organic or chemical-free.

Another Note About PVC Items:

PVC items, many of which contain obesogenic contaminants, are ubiquitous in most households. Items commonly containing PVC include vinyl flooring and blinds, wallpaper, imitation leather furniture, rain-protective coating (including rain gear), T-shirts with shiny PVC prints, mouse pads, toys, and shower curtains. PVC contains toxic organotins that are often found in house dust, along with many other toxic and obesogenic chemicals.[226,231] That quintessential smell of these soft plastics is the off-gassing of phthalates that were added for flexibility. You have to decide how much you need these items and find out if there are alternatives. Balance your risk versus benefit. You probably drive a car, which is risky, but the benefit outweighs the risk. I do not need vinyl shower curtains, window blinds, or floor coverings. When I buy baby gifts for friends, I go with natural materials such as wood and untreated fabrics. Use common sense. Make incremental changes to limit your exposure—and that of your loved ones—and you will be well on your way to a safer, cleaner, leaner environment.

OBVIOUSLY, QUIT THE HABIT

The number of people who smoke has declined in the United States in recent years, at least in part because of the settlement between the Justice Department and big tobacco companies, requiring the latter to educate the public on the dangers of smoking that were deliberately hidden in their pursuit of profits. You should also be aware that smoking still remains among the leading causes of preventable disease and death, accounting for more than 480,000 deaths every year, including more than 41,000 deaths from secondhand smoke exposure (a multimillion-dollar campaign by the tobacco industry attempted to manufacture doubt about the dangers of secondhand smoke).[217]

Overall, one of every five deaths is attributed to smoking tobacco products. The chemicals in cigarette smoke increases one's risk not only for lung cancer, but for virtually every other disease imaginable, from heart disease to dementia. On average, nonsmokers outlive smokers by ten years. If you want to really see what smoking does to your lungs, visit one of the Body Worlds exhibitions when it appears in a nearby city.[232] This exhibition uses a unique plasticization technique to reveal the intricate and beautiful details of the human body, as well as the blackened, tar-clogged lungs of habitual smokers.

While it should be obvious that smoking is not a healthy habit, few people consider the obesogenic power of this bad habit—especially when a woman smokes during pregnancy. In fact, one of the earliest links between human fetal development and obesity arose from studies of exposure to cigarette smoke in utero.[233] Although secondhand smoke exposure has decreased by more than half over the past twenty years, an estimated 40 percent of nonsmoking Americans still have nicotine by-products in their blood, showing that exposure remains widespread.[234] Now there is talk about the effects of thirdhand smoke, defined as residual nicotine and other chemicals left on a variety of indoor surfaces by tobacco smoke. This residue, which clings to and builds up on surfaces over time and is not easily wiped away by normal cleaning, is thought to react with common indoor pollutants to create a toxic, cancer-causing mix.

Do not think that if you vape, or use so-called e-cigarettes, you are reducing your risk from inhaled chemicals. At this writing, there are few, if any, regulations on what sorts of chemicals can be in vape or e-cigarettes, so you are in completely unknown, and very likely toxic, territory.

There are numerous smoking cessation programs today, including the use of drugs to wean one off the craving for nicotine. I smoked cigarettes when I was young and kicked the habit in my mid-twenties. I found that while quitting was not overly burdensome, modifying my behavior was very difficult. For me, the hardest time to avoid picking up a cigarette was when having a drink with friends. On the other hand, my dad, who was like a force of nature, was unable to quit smoking despite many attempts and died as a result of chronic obstructive pulmonary disease caused by a lifetime of smoking. Perhaps there is no greater gift to a child than a smoke-free environment beginning with conception and continuing onward.

PHARMACEUTICAL OBESOGENS IN YOUR MEDICINE CABINET

Before we move to the last chapter, let me offer some final words for those who take pharmaceutical drugs that have obesogenic side effects. As noted in chapter 5, many contemporary antidepressants and antipsychotics, for example, can trigger weight gain. While I should hope that future medicines can omit this unwanted side effect, until we have better drugs at our disposal, each of us needs to ask how badly we need these drugs and if there are any other alternatives to try that may not have obesogenic side effects. I do not want to undermine the value—and sometimes necessity—for certain drugs when warranted. But I do want people to be aware of all the potential consequences so they can make informed decisions. Doctors frequently fall far short of disclosing the complete suite of drug side effects (especially if multiple drugs are being prescribed, which is often the case). You will probably learn more about the side effects of various drugs from nightly television ads than from the few minutes you spend with your doctor, or even your pharmacist. Fortunately, you can learn a lot from the Web. I like RxList and

WebMD for their discussion of drug side effects. If your doctor prescribes a drug for you that has weight gain as a side effect, you need to ask if there is an alternative that lacks this side effect, or search it out yourself and ask the doctor whether the alternative will accomplish his goals!

CHAPTER 9

Extra Credit

Noise, Light, and Other Sources of Obesogenic Pollution Sabotaging Your Weight Loss Efforts

Now that we have discussed all of the major factors within your control, we can finish up with some broader, more environment-related topics that can limit your exposure to factors found in the environment that contribute to obesity. Your personal environment consists of where you live and work—in other words, where you spend most of your time. The "built environment" is pretty much what you guess it is—the man-made, physical parts of where we live and work, the cities, towns, roads, parks, transportation systems, all of which affect our levels of stress and physical activity. Generally speaking, the only way you can control your built environment is by choosing where to live. But you can control your personal environment by the choices you make inside your home and, perhaps to some degree, your office.

URBAN AIR POLLUTION

Los Angeles International Airport (LAX) is about forty miles from my home in Irvine (between forty minutes and four hours away, depending on traffic). When ranked by daily takeoffs and landings, LAX is the fourth-busiest airport in the world, behind Hartsfield-Jackson Atlanta International, Chicago's O'Hare International, and Dallas/Fort Worth International.[235] LAX averages more than one takeoff or landing per minute, and it is much busier during some parts of the day. My wife grew up in the Westchester neighborhood of Los Angeles, a mile or two from both the very busy 405 Freeway and LAX. When we started dating, and I experienced the house-shaking noise of a nearby takeoff for the first time, I started to wonder what health consequences were associated with living so close to major sources of noise and air pollution.

It is reasonable to think that living near a bustling airport will bring associated air-quality issues. The ultrafine sulfur dioxide, nitrogen oxide, and other toxic particles that are created from the condensation of hot exhaust vapors do not stay confined to the local area. They fan out for miles and can affect anyone breathing that air. These particles can embed themselves deep inside the lungs and then enter the bloodstream. Whether or not they are obesogenic remains to be determined conclusively, but the inflammation that they cause is suspected of worsening many lung conditions, such as asthma and chronic obstructive pulmonary disease (COPD), and of contributing to the development of heart disease. Researchers at the Keck School of Medicine at the University of Southern California have been studying the effects of air pollution on health for many years. In a 2014 paper, they showed that air pollution from LAX affects residents up to ten miles away and concluded that the impact of LAX plane traffic on air quality was so bad that "LAX should be considered one of the most important sources of [particulate matter] pollution in Los Angeles."[236] To put this into perspective, the researchers calculated that it would take between 174 and 491 miles of freeway traffic, which is about 20 to 50 percent of the highways in Los Angeles County, to match the levels of pollution detected from LAX.

A major interest of the Southern California Environmental Health Sciences Center, and its sister, the Southern California Children's Environmental Health Center, directed by my colleague Professor Rob McConnell, is in understanding the effects of the ubiquitous air pollution in the L.A. basin on the health of its residents. Center researchers have published a series of studies linking the amount of particulate air pollution in a child's local environment (for example, home and/or school neighborhood) with risk of obesity.[237] Traffic-related air pollution was linked with obesity and BMI in children aged five to eleven years.[238] They examined more than forty-five hundred children living in thirteen different communities across Southern California over four years, measuring height and weight annually, and used various statistical models to estimate and test traffic density and traffic pollution related to BMI growth. They followed the children beginning in kindergarten or first grade over the next four years and found that those exposed to the highest levels of pollution showed an almost 14 percent increase in annual BMI, and this was increased even more if the children were exposed to secondhand tobacco smoke.[239] More research is definitely needed, but what we know is that air pollution from air or street traffic is detrimental to our health and likely to increase obesity.

NOISE POLLUTION

Air pollution from bustling highways and airports is not the only culprit for increased health challenges, including obesity. Noise pollution, especially related to busy airports, has come under scrutiny lately in scientific circles. Recent studies show that people who live near airports are at increased risk for cardiovascular disease—separate from any heightened cardiovascular risks associated from the air pollution. One such study published in the *British Medical Journal* found that people who lived in the noisiest areas had an elevated risk for stroke, coronary heart disease, and cardiovascular disease, even after adjusting for confounding factors such as ethnicity, social deprivation, smoking, road traffic noise exposure, and air pollution.[240] In addition, the response to noise was dose-dependent; the risk was greatest in the 2 percent of the population who experienced the highest levels of noise. In a parallel study, noise near eighty-nine U.S. airports was significantly linked with a higher relative risk of hospital admission for cardiovascular disease in older Medicare recipients.[241]

While you might guess that exposure to constant or high levels of noise acted by interfering with sleep to impair health, this was not the case; the effects of noise were more direct. Chronic noise leads to continual stress on the body, which has repercussions such as higher blood pressure and heart rate, endocrine disruption through the stress hormone cortisol and heightened overall inflammation.[242] Dr. Robert Koch was a German physician and microbiologist who isolated the bacteria that caused anthrax and identified the bacteria responsible for cholera and tuberculosis (for which he received the Nobel Prize in Physiology or Medicine in 1905). Koch's "postulates" were four rules that must be fulfilled to prove that a microbe caused a particular disease. Koch presciently predicted in 1910 that "one day man will have to fight noise as fiercely as cholera and pest."

The World Health Organization (WHO) quantified the health burden of environmental noise in a recent report.[243] This burden was measured as "disability-adjusted life years," or the number of years lost because of disability or death. The WHO estimates that each year in Western Europeans, 45,000 years are lost through noise-induced cognitive impairment in children, 61,000 years are lost as a result of noise-induced cardiovascular disease, a whopping 903,000 years are lost owing to noise-induced sleep disturbance, and 22,000 years are lost because of tinnitus (constant ringing sound in the ears). Moreover, while not being a disease per se, noise-induced annoyance decreases quality of life and therefore also causes disability, which the WHO quantified as 587,000 disability-adjusted life years lost in the Western European population.

Now, can we call excessive, chronic noise an obesogen? While noise is not a chemical or something tangible that we can consume or inhale, we can most definitely categorize it as an obesogen—a "built-environment" obesogen. Noise is a by-product of the built environment, and it negatively affects human physiology in ways that can ultimately lead to weight gain. Although elevated stress on the body due to noise is perhaps the major mechanism here, the disruptive effects of noise on sleep are also significant. As we already discussed, sleep disturbances have numerous downstream consequences ranging from metabolic dysfunction to cardiovascular disease, depression, and immune suppression. Put simply, losing out on sleep creates a vicious biological cycle in your body, disrupting its preferred "clock" and making you more prone to many factors that contribute to weight gain. This is why getting plenty of sleep is among the suggestions in this chapter that will help you control your environmental obesogens. Let's start there.

GET A GOOD NIGHT'S SLEEP AS OFTEN AS POSSIBLE

As we discussed in chapter 5, sleep medicine has come a long way in the past decade. We know now that sleep not only has large effects on metabolism and related hormones that control weight, satiety, appetite, and cravings, but also has a big impact on circadian rhythms—our inner clock that manages those hormones and sense of day/night. The good news is that sleep is a behavior that can be reprogrammed and supported.

Our bodies crave balance and regularity, preferring to sleep on a predictable schedule. Maintaining healthy sleep habits, in fact, is what helps ensure that the body stays homeostatic and firing its hormones on cue. Millions of Americans are chronically sleep deprived and miss out on the quality of sleep that their bodies need consistently to be fully functional and healthy. Either their self-imposed behaviors (for example, staying up too late—I am guilty of this one) prevent sufficient sleep or they suffer from a real disorder that prevents them from achieving restful sleep (as with sleep apnea and insomnia). Here are some ideas about getting a better night's sleep:

Know your own needs. Everyone has different sleep needs, although most of us would do well to bank between seven and nine hours a night. One good way to determine how much sleep your body wants is to sleep in a dark room and see how long it takes for you to wake up naturally (no alarm). If you cannot do this, here is another approach to determine an optimal wake-up time. For a week, go to bed about 7.5 hours before your usual wake-up time and see if you can wake naturally five to ten minutes before your alarm goes off.[244] This number is based on the fact that the average sleep cycle is ninety minutes long and a full night's sleep includes five full cycles. You might need more or less than 7.5 hours depending on how long your individual cycles last.

Be consistent. Try to go to bed and rise at the same time daily, weekends and holidays included. Although many people try to shift their sleep habits on weekends to make up their sleep deficiency accumulated during the week, this can sabotage a healthy circadian rhythm. While it's okay to sleep an extra hour when you have the opportunity and feel like you need it, avoid waking up every morning at six thirty a.m. and then sleeping until ten on a weekend. That will throw your body clock off, and with that goes your normal metabolism and hormonal secretions.

Establish calming bedtime rituals. Try to set aside at least thirty minutes before bedtime to unwind and perform tasks that help your body know that bedtime is at hand. Disconnect from stimulating tasks (such as work, being on the computer, or using a cell phone) and engage in activities that are calming, such as taking a warm bath, reading, drinking herbal tea, or listening to soothing music. Stretch or do something relaxing.

Hit the pillow before midnight. Just as our individual sleep needs vary, so do ideal times for going to bed. The best bedtime is when you feel most sleepy before midnight. Non-REM (rapid eye movement) sleep tends to dominate sleep cycles in the early part of the night. Then, as night moves closer to dawn, dream-rich REM sleep begins to take over. Although both types of sleep are important and offer separate benefits, non-REM, slow-wave sleep is deeper and more restorative than REM sleep. Note that your ideal bedtime will likely change as you age. The older you get, the earlier your bedtime will become and the earlier you will naturally wake up.

Clean, cool, quiet, and dark. Keep your bedroom clean and cool (ideal temperature for sleeping is between 60 and 67 degrees Fahrenheit). Sleep in the dark—minimize light sources nearby, including that from an electronic device (see next tip). Consider a sleep mask if it is not possible to black out your environment. Try a sound machine or white noise generator to block out noises from the street if you live in an urban environment.

Minimize blue light from electronics. Nearly all light—whether natural sunlight in origin or artificial from light bulbs, TV screens, computers, and smartphone screens—contains blue wavelengths that are a potent suppressor of melatonin, the hormone needed for sleep, and stimulate the alert centers in the brain—a double whammy on sleep.[245] Sleep is not the only negative consequence of blue light at night. In 2015, neuroscientist Anne-Marie Chang and colleagues showed that light-emitting devices such as eReaders caused people using them to take longer to fall asleep owing to reduced sensations of sleepiness, have reduced secretion of the sleep-inducing hormone melatonin, have later timing of their circadian clock (that is, were on a "late schedule"), and be less alert the next morning than people who read paper books.[246] Modern light-emitting diodes (LEDs) produce quite a bit of blue wavelengths, and these are ubiquitous in televisions, smartphones, tablets, and computers. Blue light should be avoided for a few hours before bedtime for optimum melatonin production. Use warm wavelengths in your home LED lighting (2700K is good). If you are someone who has persistent problems falling asleep, it might be easier to get your hands on eyeglasses that filter out blue light. Make sure that your clocks, night-lights, dimmers, and so on use red, or "warm glow," lights rather than blue or green. Red light has the least power to shift circadian rhythm and suppress melatonin. Get an app that changes the color temperature of your screen to avoid blue light, particularly if you like to read in bed as I do. The one I use on my phone and tablet is called Twilight and on the computer is f.lux. Both allow me to adjust the color balance to reduce or eliminate the blue light after a certain time of night that I set.

Mind your medicines and other sources of sleep offenders. Drugs, even the ones we don't consider prototypical drugs such as coffee and alcohol, do indeed impact sleep. It can take time for the body to process caffeine, so try to have a two p.m. cutoff time if you have difficulty sleeping. Alcohol is a mixed bag with respect to sleep. While alcohol can make you feel sleepy, its effects on the body disturb normal sleep cycles and particularly disrupt the restorative slow-wave sleep. Pharmaceuticals, whether over-the-counter or prescribed, can contain ingredients that impact sleep. For example, many headache remedies contain caffeine. Some cold remedies can have stimulating decongestants (such as pseudoephedrine). Side effects in many commonly used drugs can impact sleep. Be aware of what you are taking, and if they are necessary medicines, see if you can take them earlier in the day when they will least impact sleep.

Consider a sleep study. If you have tried all of these approaches and still fail to get a good night's sleep and/or find yourself relying on sleep aids indefinitely, you might want to consider a sleep study to rule out other issues such as undiagnosed sleep apnea. This will require that you spend the night in a sleep lab that can monitor and record your sleep. These centers are not as unusual as you might think. Many hospitals large and small offer these services.

The benefits of adequate sleep extend to children as well, and they can never start too early. As with adults, kids need a consistent sleep schedule—to go to bed and wake up at the same time every single day, to maintain calming bedtime rituals, and to avoid stimulating things such as lights, screens, and sugar within bedtime.

Before moving on to the next section, I should add that exercise can also help with sleep. Generally speaking, there is a positive link between the amount of vigorous exercise you get and sleep quality.[247] Theories as to why this happens run the gamut from the stress-reducing effects of exercise (so presumably you are less likely to be keyed up about work and your "to-do list" to get to sleep on time) to other changes in body chemistry that positively impact hormonal patterns and the circadian rhythm. For example, a study by the National Sleep Foundation called the Sleep in America poll reported that older respondents (aged fifty-five to eighty-four years) who exercised more frequently had fewer complaints than those exercising less than once per week.[248] Similarly, a British study found that higher levels of regular physical activity appeared to be protective against the incidence of chronic late-life insomnia.[249] These studies demonstrate a correlation between exercise and sleep, and although they suffer from shortcomings such as the use of questionnaires, sleep diaries, and self-reported feelings of well-being, there is probably enough evidence to suggest that regular exercise should help with sleep.

GET MOVING AND KEEP MOVING

We all know that exercise is good for us for a multitude of reasons. But guess what? You do not have to train for a marathon or join a CrossFit gym. Find something you enjoy doing and do it regularly—it is easier than you might think. As with sleep medicine, exercise physiology has come a long way in my lifetime. There is no end to the number of studies showing how exercise positively impacts the body. When we decide to power walk for thirty minutes in a hilly neighborhood, go breathless in an aerobics class, or swim ten laps in a pool, our physiology changes. Some of the most eye-opening research of late has been the metabolic profiling of fit people. Not only is being physically fit associated with a healthier heart and better physique, but researchers have also identified a host of metabolic changes that occur during exercise.

A sedentary lifestyle is a known behavioral obesogen. Where people go wrong in their thinking about physical activity is that exercise alone will keep you lean (increased exercise will help keep you lean when combined with diet modification) and that you have to engage in a very rigorous routine to gain any benefits from exercise (you do not). The simple truth is that you just need to find something you enjoy so that you will do it regularly. Something that allows you to build/preserve lean muscle mass and maintain some degree of cardiorespiratory fitness. This means that your circulatory and respiratory systems are healthy enough to supply fuel and oxygen during sustained physical activity (such as exercise or sex). I will add that metabolic fitness must also be considered; you cannot be considered "fit" if you have chronic high blood sugar, insulin resistance, diabetes, high cholesterol, or any of the other obesity-related risk factors that are often lumped together and called "metabolic syndrome."

The definition of fitness has also been influenced by recent research into the effects of prolonged sitting. Studies began to emerge in 2010 showing that prolonged sitting compromises metabolic health and increases the risk of premature death—regardless of age, body weight, or amount of physical activity.[250] Peter Katzmarzyk from the Pennington Biomedical Research Center published a provocative study showing that sitting for three or more hours per day led to a two-year decrease in longevity, irrespective of overall fitness or exercise.[251] This means that an hour-long session on an exercise machine will not necessarily undo all the damage of being sedentary the rest of the day (in front of the computer, commuting, watching TV). Most of that damage is metabolic. When you are immobile, your circulation slows down and your body uses less of your blood sugar. Being motionless also negatively influences blood triglycerides, high-density lipoprotein (the good cholesterol), resting blood pressure, and the satiety hormone, leptin (which tells you when to stop eating). Sitting puts muscles into a sort of dormant state where their electrical activity is diminished. Moreover, the production of lipoprotein lipase, the enzyme that breaks down fat molecules in the blood, is shut down. And your metabolic rate plummets.

Dr. James Levine is an endocrinologist and co-director of Obesity Solutions at the Mayo Clinic and Arizona State University. He jerry-rigged what must have been one of the world's first treadmill desks in 1999 when he bought a $300 second-hand treadmill from Sears and combined it with a tray table that had a telescoping base made to fit over an elevated hospital bed. Since then he has not only promoted the power of more movement throughout the day, but published extensively on "inactivity studies"—the perils of sitting for prolonged periods. His recent work showed that fidgeting while sitting can confer benefits such as increasing energy expenditure by 20 to 30 percent (although it doesn't increase heart rate).[252]

Beyond separating the exercisers from the non-exercisers, a significant concern should be the general lack of walking, standing, and moving our bodies on a regular basis to counteract all the harm that can result from sitting for the majority of the day (hence Dr. Levine's endorsement of general fidgeting!). We know that moving routinely throughout the day—even if it is just walking around while talking on the phone, exercising while watching TV, taking the stairs instead of the elevator, using a treadmill desk for simple work tasks, or simply making a point to get up every hour for a five-minute stroll or jog in place—will have positive biological effects to counteract the poison of excessive sitting.

People also fail to appreciate just how valuable muscle mass is to quality of life and the ability to optimize metabolic health. Unlike fat, which stores calories, muscle is a highly active tissue that burns calories. This helps explain why lean, more muscular people tend to burn more calories at rest than do people with higher proportions of body fat. Numerous studies support the positive influence of muscle on a healthy metabolism. For example, researchers at UCLA found that increasing muscle mass to average or above average levels led to improved blood glucose management (that is, better insulin sensitivity and lower risk of prediabetes and diabetes).[253] In particular, the study's results established that every 10 percent increase in the ratio of skeletal muscle mass to total body weight was associated with an 11 percent reduction in the risk of insulin resistance and a 12 percent drop in the risk of developing prediabetes or diabetes.

Achieving ideal fitness is not required to improve your health. A well-rounded exercise program would be one that builds and maintains fitness and includes cardio work, strength training, and stretching. Each of these activities offers unique benefits that your body needs for peak performance and to positively affect your metabolism and longevity. Plenty of activities can cover these areas, from formal gym classes to at-home routines using streaming video over the Internet. Just be sure not to overdo it if you have not established an exercise routine already. Too much exercise will work against you and you might burn out and become a couch potato again.

For the most part, the benefits of exercise are cumulative. You can engage in short bursts of exercise throughout the day (which can actually help minimize your time spent sitting) or commit to a routine that blocks out an hour or so for your workouts. If you do dedicate a single period of time to your exercise regimen, do not allow yourself to be sedentary the rest of the day:

- Break up your sitting time by taking five- to ten-minute breaks every fifty minutes, which is around the average person's working attention span.

- Stand or exercise while watching television.
- If you work sitting down at the computer a lot, sit on something wobbly such as an exercise ball or a backless stool that forces you to engage your core muscles. Sit up straight and keep your feet flat on the floor in front of you.
- Consider a standing desk so you can alternate between sitting and standing at your workstation.
- Fidget more!

There are plenty of ingenious devices and apps (some of which you can download for free on your smartphone) to help you track your movement and exercise habits. Remember, exercise alone cannot make you lean or keep you lean. Exercise is best thought about within the context of keeping an optimal physiology and increasing your chances of living longer.

CONCLUSION

Warning: May Cause Obesity

I am fortunate to live in a state famous for its aggressive and progressive efforts to warn consumers about chemicals known to cause cancer or reproductive toxicity. In part, this stems from a ballot initiative, Proposition 65, which passed back in 1986. Prop 65 is formally known as the Safe Drinking Water and Toxic Enforcement Act of 1986. The primary state agency that enforces Prop 65 is the Office of Environmental Health Hazard Assessment (OEHHA), www.oehha.ca.gov. The main mission of OEHHA is to protect human and environmental health from the risk posed by hazardous substances using a careful scientific evaluation of these chemicals and their risks. Prop 65 requires anyone who manufactures or distributes a product sold in California that contains a material on a list of approximately eight hundred hazardous chemicals, identified by OEHHA as known to cause cancer or reproductive harm, to include a warning label. These labels and signs are now ubiquitous, posted in bars, restaurants, coffee shops, schools, and apartment buildings and on scores of products.

As you might expect, industry has fought hard to minimize

the powers of OEHHA, and industry-supported bloggers and opinionistas regularly promote disinformation, "alternative facts," and generally do their best to manufacture doubt about the scientific evidence concerning the hazards of almost any chemical that makes substantial profits for industry. As of this writing, the industry PR machine, together with the Monsanto legal team, is doing its best to convince the public and OEHHA that it is a good idea to use copious amounts of a carcinogenic chemical for cosmetic weed abatement, even around schools and parks frequented by children. When you see such an article, remember *Merchants of Doubt*, mentioned earlier[217]; doubt is their product, and if you doubt the science, you will continue to buy the product, irrespective of the potential harm. This is the cynical goal of such PR campaigns. Also as of this writing, industry has failed in one of its most recent campaigns—OEHHA has now listed glyphosate, the active ingredient in Roundup, the mostly widely used weed killer in the world, as a chemical known to the state of California to cause cancer. This was not because a few controversial studies showed this, but rather because the International Agency for Research on Cancer, a part of the World Health Organization, examined all of the published studies on glyphosate and concluded that it was a probable human carcinogen.

The Prop 65 warning labels are far from perfect, for they do not indicate what the substance is, where it is in the product, how you might be exposed to it, or the extent of the risk. They also do not offer information on how to reduce or avoid exposure. Sometimes the chemicals are man-made and avoidable and the labels obvious and somewhat redundant—does anyone really doubt that gasoline contains toxic and carcinogenic chemicals? At other times the chemicals are naturally found in foods or produced as a by-product of cooking. For example, if you are wondering why coffee shops and restaurants post Prop 65 warning signs, it is primarily because some items contain acrylamide, a substance formed in foods that are cooked at high temperatures for an extended time (as in many baked, toasted, roasted, and fried foods) by a reaction between sugars—naturally occurring or added—and a breakdown product of the amino acid asparagine. Prop 65 warning labels may also be weak because they might not always reflect the most current science—that is, they don't include as many chemicals as they could. Despite this, Prop 65 is at least a well-intended start that gives California residents peace of mind that residents of other states may not have. In response to laws such as Prop 65, the chemical industry lobbied hard for the Frank R. Lautenberg Chemical Safety for the 21st Century Act (S.697), which preempts the right of California and other states to regulate chemicals that the EPA has decided to allow. It will be very interesting to see how this plays out in the future, when California identifies a chemical under Prop 65 that the EPA allows to be used. My bet is on California. Just today (September 13, 2017), California passed State Bill 258, which will require manufacturers to disclose the ingredients in all cleaning products, destroying the "fragrance exception" we discussed in chapter 8. No doubt this will provoke a strong response from industry and its allies. I can't wait to see how this new battle turns out.

Most of us don't like to be told what to do or not do. We like freedom of choice. We do not want the government to legislate how we live so long as it does not cause harm to others and the collective good of society. But in the end, at least to some degree, we do pay a price as individuals and as a society for unhealthy behaviors such as smoking and engaging in habits that fuel the obesity epidemic, particularly when we may not know the full extent of how our behavior affects our health. Cigarette packs, in fact, were among the first products to be slapped with a warning label; in 1966 the federal government mandated that they all carry the message from the surgeon general. The initial warning was relatively mild: "Caution: Cigarette Smoking May Be Hazardous to Your Health." But the current warning is more realistic: "SURGEON GENERAL'S WARNING: Smoking Causes Lung Cancer, Heart Disease, Emphysema, and May Complicate Pregnancy." The surgeon general's warning landed on alcohol in 1989. How many more years—maybe decades—will it take for us to find similar warnings on products whose ingredients may cause obesity? And should we go this far in our labeling endeavors? I think so. But one might reasonably ask whether we should draw a line, and if so, where?

We are starting to see restaurants label their menus with calorie counts. Unfortunately, restaurants did not decide to add calorie counts because they were concerned about our health; they did so because they were mandated by law to disclose nutritional data. After all, I don't think Outback Steakhouse wants us to know that their popular Bloomin' Onion contains more than 800 calories, including 58 grams of fat (22 grams of which are saturated fat), plus 1,520 mg of sodium. And that is before you start dipping it into the sauce. Whether or not such information deters people from ordering certain foods and/or beverages to the extent it helps stem the tide of exploding obesity rates remains to be seen. It has made a difference in the foods I choose in fast-food restaurants. I applaud measures such as these because they can help us make better, more informed decisions when it comes to behaviors that impact our health.

The challenge, however, is waiting for laws that force companies to expose their practices and ingredients and the consequences of exposure to them. Let's also not forget that doctors once endorsed smoking and that the American Heart Association, in its recent endorsement of the use of polyunsaturated vegetable oils instead of the increasingly popular saturated fat coconut oil, neglected to mention the support they receive from the soy industry. Clearly, even so-called authorities can take a position that is biased and partly or completely wrong. Worse yet, there is so much information out on the Internet that it is difficult for any but the most skeptical and critical consumers to ferret out the true facts. It is up to each one of us to push for change in industry and demand more research where research is needed.

Indeed, personal choice and responsibility matter, but we are living in a world seemingly engineered to make and keep us fat (and unhealthy as well). Foods are made to be addictive. Subliminal messaging meets our eyes and brains daily, telling us that more is better and daring us to "eat just one chip." Everyday products we use expose us to chemicals that have the power to change our physiology in ways we never knew before. We take for granted the pleasures and conveniences of our modern society in the Western world, where we can order as much of anything we want at the swipe of a finger on our smartphones, but we do not think about the undertow to that modernity. Millions of us are paying a huge price. Millions more will continue to pay a hefty price in future generations if we do not do something starting today. While it was once the case that this phenomenon was most prevalent in the United States, the multinational food industry is doing a great job of exporting the obesity epidemic worldwide. Despite the prevalence of unhealthy, addictive food, we are bombarded with the message that obesity is our fault alone, that we eat too much and exercise too little and have chosen this fate. By now, I hope that the messages in this book have convinced you of the truth.

Imagine if someone took money out of your bank account without telling you. He does this at random throughout the month, and there is no paper or digital trail. When trying to balance your checkbook at the end of the month, you cannot figure out what is wrong. Nothing computes according to your data, and you cannot make sense of your checkbook. You do your best to account for income and spending and to balance the checkbook, but you continue to fail. Well, that is rather similar to what is going on today with your caloric checkbook. Yes, you can count calories in and calories burned, but you will not easily be able to factor in all the other components to your weight equation. And the presence of obesogens is a big one. That thief is robbing your personal account. Worse, that hidden, silent villain is messing with your biology as you try so hard to eat well and exercise more.

My hope is that this book has armed you with the necessary knowledge to go forward and pursue a healthier life without feeling bad about yourself, to avoid thinking that only your sloth and gluttony (two of the so-called seven deadly sins) are responsible for your ever-expanding waistline and poor health. Be the change in how you live. And be a model for others, children included.

Acknowledgments

When I started my lab at the University of California–Irvine (UCI) in 1998, I never imagined that I would work on endocrine disrupting chemicals (EDCs), obesity, or epigenetic transgenerational inheritance. However, as many successful scientists will tell you, one must follow the data where it leads, rather than trying to impose our own views on nature. My work on obesogens derives from a subtle evolution in my scientific interests that was grounded in a lifelong interest in how the embryo is patterned from the fertilized egg to a complex organism. As a postdoctoral fellow working on frog development at UCLA, I became interested in how small molecules might affect embryonic patterning and was greatly attracted to the work that Ron Evans's laboratory at the Salk Institute was doing on nuclear hormone receptors. When Ron "scooped" us on identifying the nuclear retinoid X receptor (RXR), I decided to join his lab after a little persuasion by Davo Mangelsdorf and Kaz Umesono, the two senior researchers in Ron's lab whom I had become friendly with. Ron created a very stimulating intellectual environment in his laboratory that has enabled him and his colleagues to make many, many seminal discoveries over the years. I joined Ron's lab to start seriously searching for new hormones that could activate so-called orphan nuclear receptors, receptor-like molecules for which we did not know the corresponding hormone. Ron generously let me do pretty much whatever I wanted and never complained when my interests drifted away from his. Dr. David Gardiner at UCI first got me interested in studying chemicals in the environment that could activate the retinoic acid receptor as potential causal agents for the epidemic of deformed frogs that first appeared in Minnesota. I was already identifying new ligands for retinoic acid receptors in Ron's lab so it was an easy jump to make. You can read about this story in Bill Souder's excellent book, *A Plague of Frogs*.[254] Ron did not bat an eye when I started spending a lot of effort on this project to the exclusion of what I was supposed to be working on (or when a *Nightline* news crew showed up to interview me in Ron's lab without asking to talk with him). Although I did not realize it initially, a natural intersection of developmental biology and identifying new hormone receptor ligands is the field of endocrine disruption. I remember that Ron and I were sitting together at a Keystone endocrine disruptor meeting in 1999 when he leaned over to me and said, "You could probably make a big impact in this field if you wanted to." The seed was sown.

Dave and I received a grant from the EPA to study deformed frogs after I joined the faculty at UCI and came tantalizingly close to identifying the environmental retinoid that might be responsible for the deformed frog problem. Unfortunately, the analytical methods available at the time were not up to the task and then the money ran out. But at the same time, the study of endocrine disruption was growing, particularly in Japan, and Professor Taisen (Tai) Iguchi (whom I had met at the same Keystone meeting mentioned above) invited me to an international endocrine disruptor meeting in Kobe, Japan. It was in Japan that my interest in endocrine disruptors was nurtured by the continuing interest shown by Tai and Professors Jun Kanno, Tohru

Inoue, Masami Muramatsu, Satoshi Inoue, Yoshi Nagahama, and others. Their many invitations to speak about our work in Japan kept me abreast on what was happening in the field and convinced me to keep working in this area before our exciting discovery of obesogens. It was in Japan that I first met Professor Howard Bern, from UC Berkeley, who had a great influence on my career. Howard believed that the measure of one's scientific career is not in what we discovered, but in "who we left behind," that is, the achievements of the people we trained. This had a profound impact on my thinking. I "adopted" Howard as my wise grandfather since I never knew either of my actual grandfathers.

Professors John McLachlan from Tulane University and Lou Guillette from the University of Florida were also major influences on my developing interest in endocrine disruptors and as role models. Working together with the global community of endocrine disruptor researchers has been a major highlight of my scientific career. It gives me great pleasure and satisfaction that the work we do has direct impact on human health. Rather than the sometimes cutthroat competition that I faced in other fields, the endocrine disruptor community was welcoming and supportive, and readily accommodated new ideas and approaches. In addition to those I have already mentioned, colleagues and collaborators I wish to thank for their friendship, collaboration, and inspiration over the years include Tom Zoeller, Fred vom Saal, Laura Vandenberg, Chiharu Tohyama, Ana Soto, Toshi Shioda, Pete Myers, Lars and Monica Lind, Juliette Legler, Jerry Heindel, Andrea Gore, David Crews, and Andrés Carrasco. I thank Drs. Mike Skinner and John McCarrey for many stimulating discussions about epigenetic inheritance, most of which took place over Tusker beers on safaris in Kenya. Many thanks also to my green chemistry colleagues,

Terry Collins and John Warner, who have totally changed how I think about chemical synthesis and design. Terry and John are truly changing the world. I hope that anyone I may have missed can forgive my leaky memory. Sadly Howard, Lou, and Andrés are no longer with us, but their influence on those of us who knew them is eternal.

The obesogen story has been a wild ride from its initial nearly absolute dismissal by the biomedical research community to the now growing acceptance that chemical effects on the genome and epigenome contribute to many chronic diseases, including obesity. Tai deserves the credit (or blame) for my interest in tributyltin (TBT) since it was through a collaboration with him that my laboratory first began to test this chemical, and it was at a meeting in Matsuyama, Japan, to which Tai invited me, where I first heard about the effects of TBT on vertebrates. I thank the NIEHS for its continuing financial support of the research in my laboratory, as well as the EPA and the greatly missed California Toxic Substances Research and Training Program, which was the first organization to find this work worthy of financial support.

Many people in my laboratory have worked on aspects of this project over the years. Dr. Raquel Chamorro-García has spearheaded all of the tedious and time-consuming transgenerational studies that are revolutionizing what we know about the effects of EDCs on physiology and the epigenome. Dr. Carlos Díaz-Castillo and my genomic collaborator Dr. Toshi Shioda from Massachusetts General Hospital have played important roles in helping to unravel what is happening to the epigenome as a result of ancestral TBT exposure. Soon-to-be Dr. Bassem Shoucri made major advances in figuring out just how EDCs affect the commitment of mesenchymal stem cells to the fat lineage through RXR, probably as a dimeric complex containing both RXR and PPARγ. Dr. Séverine Kirchner first found that TBT caused effects on the epigenome that altered the fate of mesenchymal stem cells. Séverine, Dr. Jasmine Li, and graduate student John Ycaza showed that the ability of TBT to change mesenchymal stem cells into fat cells required action of the nuclear receptor, PPARγ. Dr. Felix Grün was the first person in my lab to work on TBT and it was during a conversation about what to call the effects of chemicals like TBT on cells and animals that we coined the term "obesogen." Many undergraduate students, graduate students, visiting scientists, and technicians contributed to this project over the years, including Rachelle Abbey, Tim Abreo, Sathya Balachander, Christie Boulos, Connie Chow, Giorgio Dimastrogiovanni, Heidi Käch, Tiffany Kieu, Jhyme Laude, Ron Leavitt, Lauren Maeda, Eric Martinez, Nina Ngyuen, Hang Pham, Nhieu Pham, Margaret Sahu, Weiyi Tang, Camilla Taxvig, Lenka Vanek, and Zamaneh Zamanian. A very special thanks is due to Dr. Amanda Janesick who as a graduate student in my laboratory undertook the writing of many reviews about the effects of obesogens with me, despite that this was not the topic of her thesis research. Amanda and I spent thousands of hours discussing and debating obesity and obesogens together and with other endocrine disruptor researchers around the world. Amanda's irrepressible enthusiasm as a writing partner made working on these many reviews enjoyable and they ultimately served as the backbone of this book.

The genesis of this book was much faster. In January 2013, I received a call from the *New York Times* columnist Nicholas Kristof who had decided to write a story about obesogens. After a pleasant conversation and a few e-mails, Nick wrote an article that appeared in the Sunday Review on January 19, 2013, entitled "Warnings from a Flabby Mouse."[255] Many thanks to Nick for his interest and for writing the excellent article that started the chain of events culminating in this book. His article attracted the attention of several literary agents who later contacted me. Bonnie Solow, who is now my amazing agent, gently persuaded me that it was time to write a book about obesogens and that I was the one to write it. After many phone calls with Bonnie, I agreed, but spent the next two years proving convincingly that I could not find the time necessary to write the required book proposal. Undaunted, Bonnie connected me with collaborator Kristin Loberg. Kristin and I clicked immediately, worked marvelously together, and became fast friends. She did a masterful job of converting our many conversations and my published papers, talks, and random musings into the skeleton of this book. Throughout countless drafts bounced back and forth between us, Kristin helped me strike a good balance between being too technical and too trivial. Together we produced something I think is much better than what either of us could have done alone. Kudos also to Karen Murgolo and her staff at Grand Central Life & Style who have helped immensely in the evolution of this book. I have used the cartoons of Eric Jay Decetis and Geoff Olson to good effect in my talks for years and am thankful to both for permission to include these cartoons in this book. Thanks also to Jim Janesick for his illustration of the somatotypes and to Retha Newbold for the photograph of her flabby mice that started the field. I am also grateful to Tom Zoeller for agreeing to let me use his provocative quotes and anecdotes. Dr. Jerry Heindel read every word of the book and offered many helpful suggestions. Dr. Peter Turnbaugh provided detailed commentary on the microbiome section. Any errors and omissions remaining in the book are entirely mine.

I want to specially acknowledge Dr. Jerry Heindel whose optimism and encouragement helped get me through a very difficult time at the dawn of obesogen research. As the NIH program officer assigned to my TBT grants, Jerry was responsible for explaining to me why the study sections that reviewed my grants didn't think they were worth funding. I remember he once said to me, "Listen, I don't have any funds to give you, but I think that this work is really, really important and I hope you don't give up, no matter how bad these reviews seem. I am sure there is something here and you are the one to find it." I was a newly minted associate professor at that time just starting to work on a controversial new topic. This strong encouragement from an NIH official helped to keep my enthusiasm high until we finally had so much data in support of our model that the grant was funded. Jerry has become a good friend over the years and we have also spent countless hours discussing obesogens, metabolism, and endocrine disrupters. Jerry was a great travel partner and didn't mind that my fast driving made him carsick. He is always up for talking science over a beer or two and his insights and honesty (occasionally brutal) have helped to hone my ideas and approaches.

Last, but not least, I thank my wife, Dejoie, and my daughter, Arielle, for their constant support and not complaining about my frequent absences over the years. I was fortunate to be invited to speak about our endocrine disruptor and obesogen research around the world and to occasionally take them with me to exotic new destinations. I was even luckier still that they entertained each other while I was abroad or when I abandoned them to attend scientific meetings in a strange city where they didn't speak the local language. Arielle was blessed to grow up with a large extended family of scientific "aunts and uncles" who doted on her even more than we did. You know who you are.

Acknowledgments

This book grew from work in my laboratory performed between 2003 and 2017, together with the work of colleagues in the field of endocrine disruption and in other disciplines. It reflects the evolution of my thinking and that in related fields over the same time period. The idea that chemicals might cause or contribute to obesity was very "counter-culture" when we started so the narrative needed to be built from scratch. It was my original intent to write a book focused on the chemical obesogens that my lab studies, but it became clear during the writing that a different type of story was needed—a modern explanation of why and how we become fat, despite trying hard not to. I have tried to incorporate the most up-to-date evidence from all of the fields that we currently know may be related to obesity and to present the most complete and accessible story possible in the hope that it may help you to gain control of your weight and health. I hope that you enjoy the ride.

References

1. Grün F, Blumberg B. Environmental obesogens: organotins and endocrine disruption via nuclear receptor signaling. *Endocrinology.* 2006;147(suppl 6):S50-S55.
2. Lind L, Lind PM, Lejonklou MH, et al. Uppsala consensus statement on environmental contaminants and the global obesity epidemic. *Environ Health Perspect.* 2016;124(5):A81-A83.
3. Keys A, Fidanza F, Karvonen MJ, Kimura N, Taylor HL. Indices of relative weight and obesity. *J Chronic Dis.* 1972;25(6):329-343.
4. World Health Organization. *Obesity: Preventing and Managing the Global Epidemic.* Geneva, Switzerland: World Health Organization; 2000. WHO Technical Report Series 894.
5. Brown RE, Sharma AM, Ardern CI, Mirdamadi P, Kuk JL. Secular differences in the association between caloric intake, macronutrient intake, and physical activity with obesity. *Obes Res Clin Pract.* 2016;10(3):243-255.
6. World Health Organization. *Report of the Commission on Ending Childhood Obesity.* Geneva, Switzerland: World Health Organization; 2016.
7. Heindel JJ, Blumberg B, Cave M, et al. Metabolism disrupting chemicals and metabolic disorders. *Reprod Toxicol.* 2017;68: 3-33.
8. Janesick AS, Schug TT, Heindel JJ, Blumberg B. Environmental chemicals and obesity. In: Bray G, Bouchard C, eds. *Epidemiology, Etiology and Physiopathology.* Boca Raton, FL: CRC Press; 2014:471-488. *Handbook of Obesity*; vol. 1.

9. National Institute of Diabetes and Digestive and Kidney Diseases. Overweight & obesity statistics. https://www.niddk.nih.gov/health-information/health-statistics/Pages/overweight-obesity-statistics.aspx. Published 2014. Accessed May 4, 2017.
10. Flegal KM, Carroll MD, Kit BK, Ogden CL. Prevalence of obesity and trends in the distribution of body mass index among US adults, 1999-2010. *JAMA*. 2012;307(5):491-497.
11. Flegal KM, Kruszon-Moran D, Carroll MD, Fryar CD, Ogden CL. Trends in obesity among adults in the United States, 2005 to 2014. *JAMA*. 2016;315(21):2284-2291.
12. CDC/National Center for Health Statistics. Obesity and overweight. https://www.cdc.gov/nchs/fastats/obesity-overweight.htm. Published 2016. Accessed May 15, 2017.
13. National Center for Health Statistics. *Health, United States, 2015: With Special Feature on Racial and Ethnic Health Disparities*. 2016 ed. Hyattsville, MD: Centers for Disease Control and Prevention; 2015:461.
14. Klimentidis YC, Beasley TM, Lin HY, et al. Canaries in the coal mine: a cross-species analysis of the plurality of obesity epidemics. *Proceedings of the Royal Society B: Biological Sciences*. 2011;278(1712):1626-1632.
15. Johnson PM, Kenny PJ. Dopamine D2 receptors in addiction-like reward dysfunction and compulsive eating in obese rats. *Nat Neurosci*. 2010;13(5):635-641.
16. Sankararaman S, Mallick S, Patterson N, Reich D. The combined landscape of Denisovan and Neanderthal ancestry in present-day humans. *Curr Biol*. 2016;26(9):1241-1247.
17. Neel JV. Diabetes mellitus: a "thrifty" genotype rendered detrimental by "progress"? *Am J Hum Genet*. 1962;14:353-362.
18. Wang G, Speakman JR. Analysis of positive selection at single nucleotide polymorphisms associated with body mass index does not support the "thrifty gene" hypothesis. *Cell Metab*. 2016;24(4):531-541.
19. Zeevi D, Korem T, Zmora N, et al. Personalized nutrition by prediction of glycemic responses. *Cell*. 2015;163(5):1079-1094.

20. Hermans MP, Amoussou-Guenou KD, Bouenizabila E, Sadikot SS, Ahn SA, Rousseau MF. The normal-weight type 2 diabetes phenotype revisited. *Diabetes Metab Syndr.* 2016;10(2 suppl 1): S82-S88.
21. Sheldon W. *The Varieties of Human Physique: An Introduction to Constitutional Psychology.* New York: Harper & Bros; 1940.
22. Sheldon WH, Dupertuis CW, McDermott E. Atlas of men: a guide for somatotyping the adult male at all ages. *J Am Med Assoc.* 1954;156(13):1294-1295.
23. Janesick A, Blumberg B. Endocrine disrupting chemicals and the developmental programming of adipogenesis and obesity. *Birth Defects Res C Embryo Today.* 2011;93(1):34-50.
24. O'Rahilly S, Farooqi IS. The genetics of obesity in humans. In: De Groot LJ, Chrousos G, Dungan K, et al., eds. *Endotext.* South Dartmouth, MA: MDText.com; 2000.
25. Locke AE, Kahali B, Berndt SI, et al. Genetic studies of body mass index yield new insights for obesity biology. *Nature.* 2015;518(7538):197-206.
26. Larder R, Lim CT, Coll AP. Genetic aspects of human obesity. *Handb Clin Neurol.* 2014;124:93-106.
27. Manco M, Dallapiccola B. Genetics of pediatric obesity. *Pediatrics.* 2012;130(1):123-133.
28. Physical Activity Council. The Physical Activity Council's annual study tracking sports, fitness, and recreation participation in the US. 2017 Participation Report. http://www.physicalactivitycouncil.com/pdfs/current.pdf. Published March 30, 2017. Accessed May 10, 2017.
29. Sawyer BJ, Bhammar DM, Angadi SS, et al. Predictors of fat mass changes in response to aerobic exercise training in women. *J Strength Cond Res.* 2015;29(2):297-304.
30. Melanson EL, Keadle SK, Donnelly JE, Braun B, King NA. Resistance to exercise-induced weight loss: compensatory behavioral adaptations. *Med Sci Sports Exerc.* 2013;45(8):1600-1609.
31. Crabtree DR, Chambers ES, Hardwick RM, Blannin AK. The effects of high-intensity exercise on neural responses to images of food. *Am J Clin Nutr.* 2014;99(2):258-267.

32. Finlayson G, Caudwell P, Gibbons C, Hopkins M, King N, Blundell J. Low fat loss response after medium-term supervised exercise in obese is associated with exercise-induced increase in food reward. *J Obes.* 2011;2011:615624. doi:10.1155/2011/615624.
33. Evero N, Hackett LC, Clark RD, Phelan S, Hagobian TA. Aerobic exercise reduces neuronal responses in food reward brain regions. *J Appl Physiol (1985).* 2012;112(9):1612-1619.
34. Singh M. Mood, food, and obesity. *Front Psychol.* 2014;5:925.
35. Alonso-Alonso M, Woods SC, Pelchat M, et al. Food reward system: current perspectives and future research needs. *Nutr Rev.* 2015;73(5):296-307.
36. United States Senate Select Committee on Nutrition and Human Needs. *Dietary Goals for the United States.* Washington, DC: US Government Printing Office; 1977.
37. Kitada K, Daub S, Zhang Y, et al. High salt intake reprioritizes osmolyte and energy metabolism for body fluid conservation. *J Clin Invest.* 2017;127(5):1944-1959.
38. Fildes A, Charlton J, Rudisill C, Littlejohns P, Prevost AT, Gulliford MC. Probability of an obese person attaining normal body weight: cohort study using electronic health records. *Am J Public Health.* 2015:105(9):e54-e59.
39. Kraschnewski JL, Boan J, Esposito J, et al. Long-term weight loss maintenance in the United States. *Int J Obes.* 2010;34(11): 1644-1654.
40. Baillie-Hamilton PF. Chemical toxins: a hypothesis to explain the global obesity epidemic. *J Altern Complement Med.* 2002;8(2): 185-192.
41. Heindel JJ. Endocrine disruptors and the obesity epidemic. *Toxicol Sci.* 2003;76(2):247-249.
42. Grun F, Watanabe H, Zamanian Z, et al. Endocrine-disrupting organotin compounds are potent inducers of adipogenesis in vertebrates. *Mol Endocrinol.* 2006;20(9):2141-2155.
43. Diamanti-Kandarakis E, Bourguignon JP, Giudice LC, et al. Endocrine-disrupting chemicals: an Endocrine Society scientific statement. *Endocr Rev.* 2009;30(4):293-342.

44. Gore AC, Chappell VA, Fenton SE, et al. EDC-2: the Endocrine Society's second scientific statement on endocrine-disrupting chemicals. *Endocr Rev.* 2015;36(6):E1-E150.
45. Weiss B. The intersection of neurotoxicology and endocrine disruption. *Neurotoxicology.* 2012;33(6):1410-1419.
46. Mendez MA, Garcia-Esteban R, Guxens M, et al. Prenatal organochlorine compound exposure, rapid weight gain, and overweight in infancy. *Environ Health Perspect.* 2011;119(2):272-278.
47. Tang-Péronard JL, Heitmann BL, Andersen HR, et al. Association between prenatal polychlorinated biphenyl exposure and obesity development at ages 5 and 7 y: a prospective cohort study of 656 children from the Faroe Islands. *Am J Clin Nutr.* 2014;99(1):5-13.
48. Valvi D, Mendez MA, Martinez D, et al. Prenatal concentrations of polychlorinated biphenyls, DDE, and DDT and overweight in children: a prospective birth cohort study. *Environ Health Perspect.* 2012;120(3):451-457.
49. Newbold RR, Padilla-Banks E, Jefferson WN. Environmental estrogens and obesity. *Mol Cell Endocrinol.* 2009;304(1-2):84-89.
50. Hines EP, White SS, Stanko JP, Gibbs-Flournoy EA, Lau C, Fenton SE. Phenotypic dichotomy following developmental exposure to perfluorooctanoic acid (PFOA) in female CD-1 mice: low doses induce elevated serum leptin and insulin, and overweight in mid-life. *Mol Cell Endocrinol.* 2009;304(1-2):97-105.
51. Newbold RR, Padilla-Banks E, Snyder RJ, Jefferson WN. Perinatal exposure to environmental estrogens and the development of obesity. *Mol Nutr Food Res.* 2007;51(7):912-917.
52. Bunyan J, Murrell EA, Shah PP. The induction of obesity in rodents by means of monosodium glutamate. *Br J Nutr.* 1976;35(1):25-39.
53. Walker RW, Dumke KA, Goran MI. Fructose content in popular beverages made with and without high-fructose corn syrup. *Nutrition.* 2014;30(7-8):928-935.
54. Goran MI, Martin AA, Alderete TL, Fujiwara H, Fields DA. Fructose in breast milk is positively associated with infant body composition at 6 months of age. *Nutrients.* 2017;9(2). pii: E146.

55. Attina TM, Hauser R, Sathyanarayana S, et al. Exposure to endocrine-disrupting chemicals in the USA: a population-based disease burden and cost analysis. *Lancet Diabetes Endocrinol.* 2016;4(12):996-1003.
56. Trasande L, Zoeller RT, Hass U, et al. Estimating burden and disease costs of exposure to endocrine-disrupting chemicals in the European Union. *J Clin Endocrinol Metab.* 2015;100(4):1245-1255.
57. Fothergill E, Guo J, Howard L, et al. Persistent metabolic adaptation 6 years after "The Biggest Loser" competition. *Obesity.* 2016;24(8):1612-1619.
58. Bennett W, Gurin J. *The Dieter's Dilemma: Eating Less and Weighing More.* New York: Basic Books; 1982.
59. Müller MJ, Bosy-Westphal A, Heymsfield SB. Is there evidence for a set point that regulates human body weight? *F1000 Medicine Reports.* 2010;2:59.
60. Spalding KL, Arner E, Westermark PO, et al. Dynamics of fat cell turnover in humans. *Nature.* 2008;453(7196):783-787.
61. Mauer MM, Harris RB, Bartness TJ. The regulation of total body fat: lessons learned from lipectomy studies. *Neurosci Biobehav Rev.* 2001;25(1):15-28.
62. Reyne Y, Nougues J, Vezinhet A. Adipose tissue regeneration in 6-month-old and adult rabbits following lipectomy. *Proc Soc Exp Biol Med.* 1983;174(2):258-264.
63. Zhang Y, Proenca R, Maffei M, Barone M, Leopold L, Friedman JM. Positional cloning of the mouse obese gene and its human homologue. *Nature.* 1994;372(6505):425-432.
64. Tontonoz P, Hu E, Spiegelman BM. Stimulation of adipogenesis in fibroblasts by PPAR gamma 2, a lipid-activated transcription factor. *Cell.* 1994;79(7):1147-1156.
65. Simpson ER. Sources of estrogen and their importance. *J Steroid Biochem Mol Biol.* 2003;86(3-5):225-230.
66. Ratajczak MZ, Bartke A, Darzynkiewicz Z. Prolonged growth hormone/insulin/insulin-like growth factor nutrient response signaling pathway as a silent killer of stem cells and a culprit in aging. *Stem Cell Rev.* 2017;13(4):443-453.

67. Harms M, Seale P. Brown and beige fat: development, function and therapeutic potential. *Nat Med.* 2013;19(10):1252-1263.
68. Rosen ED, Spiegelman BM. What we talk about when we talk about fat. *Cell.* 2014;156(1-2):20-44.
69. Wu J, Bostrom P, Sparks LM, et al. Beige adipocytes are a distinct type of thermogenic fat cell in mouse and human. *Cell.* 2012;150(2):366-376.
70. Berry DC, Jiang Y, Graff JM. Emerging roles of adipose progenitor cells in tissue development, homeostasis, expansion and thermogenesis. *Trends Endocrinol Metab.* 2016;27(8):574-585.
71. Phillips CM, Perry IJ. Does inflammation determine metabolic health status in obese and nonobese adults? *J Clin Endocrinol Metab.* 2013;98(10):E1610-E1619.
72. Sahakyan KR, Somers VK, Rodriguez-Escudero JP, et al. Normal-weight central obesity: implications for total and cardiovascular mortality. *Ann Intern Med.* 2015;163(11):827-835.
73. Berry DC, Stenesen D, Zeve D, Graff JM. The developmental origins of adipose tissue. *Development.* 2013;140(19):3939-3949.
74. Wei H, Averill MM, McMillen TS, et al. Modulation of adipose tissue lipolysis and body weight by high-density lipoproteins in mice. *Nutr Diabetes.* 2014;4:e108.
75. Taubes G. *Why We Get Fat: And What to Do About It.* New York: Knopf; 2010.
76. Tontonoz P, Spiegelman BM. Fat and beyond: the diverse biology of PPARgamma. *Annu Rev Biochem.* 2008;77:289-312.
77. Chamorro-García R, Diaz-Castillo C, Shoucri BM, et al. Ancestral perinatal obesogen exposure results in a transgenerational thrifty phenotype in mice. *Nat Commun.* 2017;in press.
78. Colborn T, Dumanoski D, Myers JP. *Our Stolen Future: Are We Threatening Our Fertility, Intelligence, and Survival?: A Scientific Detective Story.* New York: Dutton; 1996.
79. Zoeller RT, Brown TR, Doan LL, et al. Endocrine-disrupting chemicals and public health protection: a statement of principles from the Endocrine Society. *Endocrinology.* 2012;153(9):4097-4110.
80. Bergman Å, Heindel JJ, Jobling S, Kidd KA, Zoeller RT, eds. *State of the Science of Endocrine Disrupting Chemicals—2012.*

Geneva, Switzerland: United Nations Environment Programme, World Health Organization; 2013.

81. Carson R. *Silent Spring.* Boston, MA: Houghton Mifflin; Cambridge, MA: Riverside Press; 1962.
82. Smith R, Lourie B, Dopp S. *Slow Death by Rubber Duck: How the Toxic Chemistry of Everyday Life Affects Our Health.* 1st ed. Toronto: A. A. Knopf Canada; 2009.
83. Borzelleca JF. Paracelsus: herald of modern toxicology. *Toxicol Sci.* 2000;53(1):2-4.
84. Herbst AL, Bern HA. *Developmental Effects of Diethylstilbestrol (DES) in Pregnancy.* New York: Thieme-Stratton; 1981.
85. Newbold RR, Padilla-Banks E, Snyder RJ, Jefferson WN. Developmental exposure to estrogenic compounds and obesity. *Birth Defects Res A Clin Mol Teratol.* 2005;73(7):478-480.
86. Rantakokko P, Main KM, Wohlfart-Veje C, et al. Association of placenta organotin concentrations with growth and ponderal index in 110 newborn boys from Finland during the first 18 months of life: a cohort study. *Environ Health.* 2014;13(1):45.
87. Li X, Pham HT, Janesick AS, Blumberg B. Triflumizole is an obesogen in mice that acts through peroxisome proliferator activated receptor gamma (PPARgamma). *Environ Health Perspect.* 2012;120(12):1720-1726.
88. Chamorro-García R, Kirchner S, Li X, et al. Bisphenol A diglycidyl ether induces adipogenic differentiation of multipotent stromal stem cells through a peroxisome proliferator-activated receptor gamma-independent mechanism. *Environ Health Perspect.* 2012;120(7):984-989.
89. Janesick AS, Dimastrogiovanni G, Vanek L, et al. On the utility of ToxCast and ToxPi as methods for identifying new obesogens. *Environ Health Perspect.* 2016;124(8):1214-1226.
90. Chamorro-García R, Sahu M, Abbey RJ, Laude J, Pham N, Blumberg B. Transgenerational inheritance of increased fat depot size, stem cell reprogramming, and hepatic steatosis elicited by prenatal exposure to the obesogen tributyltin in mice. *Environ Health Perspect.* 2013;121(3):359-366.

91. Anway MD, Cupp AS, Uzumcu M, Skinner MK. Epigenetic transgenerational actions of endocrine disruptors and male fertility. *Science.* 2005;308(5727):1466-1469.
92. Skinner MK. Environmental epigenetics and a unified theory of the molecular aspects of evolution: a neo-Lamarckian concept that facilitates neo-Darwinian evolution. *Genome Biol Evol.* 2015;7(5):1296-1302.
93. Manikkam M, Tracey R, Guerrero-Bosagna C, Skinner MK. Plastics derived endocrine disruptors (BPA, DEHP and DBP) induce epigenetic transgenerational inheritance of obesity, reproductive disease and sperm epimutations. *PLoS One.* 2013;8(1):e55387.
94. Tracey R, Manikkam M, Guerrero-Bosagna C, Skinner MK. Hydrocarbons (jet fuel JP-8) induce epigenetic transgenerational inheritance of obesity, reproductive disease and sperm epimutations. *Reprod Toxicol.* 2013;36:104-116.
95. Skinner MK, Manikkam M, Tracey R, Guerrero-Bosagna C, Haque M, Nilsson EE. Ancestral dichlorodiphenyltrichloroethane (DDT) exposure promotes epigenetic transgenerational inheritance of obesity. *BMC Med.* 2013;11:228.
96. Li DK, Miao M, Zhou Z, et al. Urine bisphenol-A level in relation to obesity and overweight in school-age children. *PLoS One.* 2013;8(6):e65399.
97. Kanno J. Introduction to the concept of signal toxicity. *J Toxicol Sci.* 2016;41(special):SP105-SP109.
98. Uribarri J, Woodruff S, Goodman S, et al. Advanced glycation end products in foods and a practical guide to their reduction in the diet. *J Am Diet Assoc.* 2010;110(6):911-916.e12.
99. Taheri S, Lin L, Austin D, Young T, Mignot E. Short sleep duration is associated with reduced leptin, elevated ghrelin, and increased body mass index. *PLoS Med.* 2004;1(3):e62.
100. La Merrill M, Karey E, Moshier E, et al. Perinatal exposure of mice to the pesticide DDT impairs energy expenditure and metabolism in adult female offspring. *PLoS One.* 2014;9(7):e103337.

101. Jeffery E, Church CD, Holtrup B, Colman L, Rodeheffer MS. Rapid depot-specific activation of adipocyte precursor cells at the onset of obesity. *Nat Cell Biol.* 2015;17(4):376-385.
102. Kaiser J. NIH cancels massive U.S. children's study. *Science Magazine: ScienceInsider*: American Association for the Advancement of Science; December 12, 2014. Accessed May 11, 2017.
103. Judson HF. *The Eighth Day of Creation: Makers of the Revolution in Biology.* New York: Simon & Schuster; 1979.
104. Watson JD. *The Double Helix: A Personal Account of the Discovery of the Structure of DNA.* 1st ed. New York: Atheneum; 1968.
105. Crick F. *What Mad Pursuit: A Personal View of Scientific Discovery.* New York: Basic Books; 1988.
106. Watson JD, Crick FH. Molecular structure of nucleic acids: a structure for deoxyribose nucleic acid. *Nature.* 1953;171(4356): 737-738.
107. Meselson M, Stahl FW. The replication of DNA in *Escherichia coli. Proc Natl Acad Sci U S A.* 1958;44(7):671-682.
108. GBD 2015 Risk Factors Collaborators. Global, regional, and national comparative risk assessment of 79 behavioural, environmental and occupational, and metabolic risks or clusters of risks, 1990-2015: a systematic analysis for the Global Burden of Disease Study 2015. *Lancet.* 2016;388(10053):1659-1724.
109. World Health Organization. Global Health Observatory (GHO) data: NCD mortality and morbidity. http://www.who.int/gho/ncd/mortality_morbidity/en. Published 2016. Accessed June 20, 2017.
110. Xu J, Murphy SL, Kochanek KD, Arias E. Mortality in the United States, 2015. *NCHS Data Brief.* 2016(267):1-8.
111. Clayton P, Rowbotham J. How the mid-Victorians worked, ate and died. *Int J Environ Res Public Health.* 2009;6(3):1235-1253.
112. Gould SJ, Eldredge N. Punctuated equilibria: the tempo and mode of evolution reconsidered. *Paleobiology.* 1977;3:115-151.
113. Wolstenholme JT, Edwards M, Shetty SR, et al. Gestational exposure to bisphenol A produces transgenerational changes in behaviors and gene expression. *Endocrinology.* 2012;153(8):3828-3838.

114. Sarkar D, Leung EY, Baguley BC, Finlay GJ, Askarian-Amiri ME. Epigenetic regulation in human melanoma: past and future. *Epigenetics*. 2015;10(2):103-121.
115. Montes M, Nielsen MM, Maglieri G, et al. The lncRNA MIR31HG regulates p16(INK4A) expression to modulate senescence. *Nat Commun*. 2015;6:6967.
116. Barker DJ, Winter PD, Osmond C, Margetts B, Simmonds SJ. Weight in infancy and death from ischaemic heart disease. *Lancet*. 1989;2(8663):577-580.
117. Forsdahl A. Are poor living conditions in childhood and adolescence an important risk factor for arteriosclerotic heart disease? *Br J Prev Soc Med*. 1977;31(2):91-95.
118. Gluckman PD, Hanson MA. Living with the past: evolution, development, and patterns of disease. *Science*. 2004;305(5691): 1733-1736.
119. Gluckman PD, Hanson MA. *Mismatch: Why Our World No Longer Fits Our Bodies*. Oxford; New York: Oxford University Press; 2006.
120. Lenz W, Knapp K. Thalidomide embryopathy. *Arch Environ Health*. 1962;5:100-105.
121. Painter RC, de Rooij SR, Bossuyt PM, et al. Maternal nutrition during gestation and carotid arterial compliance in the adult offspring: the Dutch famine birth cohort. *J Hypertens*. 2007;25(3): 533-540.
122. Heijmans BT, Tobi EW, Stein AD, et al. Persistent epigenetic differences associated with prenatal exposure to famine in humans. *Proc Natl Acad Sci U S A*. 2008;105(44):17046-17049.
123. Pembrey ME, Bygren LO, Kaati G, et al. Sex-specific, male-line transgenerational responses in humans. *Eur J Hum Genet*. 2006;14(2):159-166.
124. Gordon JI. Honor thy gut symbionts redux. *Science*. 2012;336 (6086):1251-1253.
125. Xu J, Gordon JI. Honor thy symbionts. *Proc Natl Acad Sci U S A*. 2003;100(18):10452-10459.
126. Conlon MA, Bird AR. The impact of diet and lifestyle on gut microbiota and human health. *Nutrients*. 2014;7(1):17-44.

127. Turnbaugh PJ, Ley RE, Hamady M, Fraser-Liggett CM, Knight R, Gordon JI. The human microbiome project. *Nature.* 2007;449(7164):804-810.
128. Blaser MJ, Dominguez-Bello MG. The human microbiome before birth. *Cell Host Microbe.* 2016;20(5):558-560.
129. Bokulich NA, Chung J, Battaglia T, et al. Antibiotics, birth mode, and diet shape microbiome maturation during early life. *Sci Transl Med.* 2016;8(343):343ra382.
130. Turnbaugh PJ, Ley RE, Mahowald MA, Magrini V, Mardis ER, Gordon JI. An obesity-associated gut microbiome with increased capacity for energy harvest. *Nature.* 2006;444(7122): 1027-1031.
131. Ridaura VK, Faith JJ, Rey FE, et al. Gut microbiota from twins discordant for obesity modulate metabolism in mice. *Science.* 2013;341(6150):1241214.
132. Caesar R, Tremaroli V, Kovatcheva-Datchary P, Cani PD, Backhed F. Crosstalk between gut microbiota and dietary lipids aggravates WAT inflammation through TLR signaling. *Cell Metab.* 2015;22(4):658-668.
133. Alcock J, Maley CC, Aktipis CA. Is eating behavior manipulated by the gastrointestinal microbiota? Evolutionary pressures and potential mechanisms. *Bioessays.* 2014;36(10):940-949.
134. Liou AP, Paziuk M, Luevano JM, Jr., Machineni S, Turnbaugh PJ, Kaplan LM. Conserved shifts in the gut microbiota due to gastric bypass reduce host weight and adiposity. *Sci Transl Med.* 2013;5(178):178ra141.
135. Tremaroli V, Karlsson F, Werling M, et al. Roux-en-Y gastric bypass and vertical banded gastroplasty induce long-term changes on the human gut microbiome contributing to fat mass regulation. *Cell Metab.* 2015;22(2):228-238.
136. US Department of Agriculture Economic Research Service. Food availability and consumption. https://www.ers.usda.gov/data-products/ag-and-food-statistics-charting-the-essentials/food-availability-and-consumption. Updated October 18, 2016. Accessed May 12, 2017.

137. Lustig RH. *Fat Chance: Beating the Odds Against Sugar, Processed Food, Obesity, and Disease.* New York: Hudson Street Press; 2012.
138. Ventura EE, Davis JN, Goran MI. Sugar content of popular sweetened beverages based on objective laboratory analysis: focus on fructose content. *Obesity.* 2011;19(4):868-874.
139. Payne AN, Chassard C, Lacroix C. Gut microbial adaptation to dietary consumption of fructose, artificial sweeteners and sugar alcohols: implications for host-microbe interactions contributing to obesity. *Obes Rev.* 2012;13(9):799-809.
140. Goran MI. My open letter to First Lady Michelle Obama regarding her comments on high-fructose corn syrup. http://www.goranlab.com/news/LettertoMObamaApril2014.pdf. Published April 30, 2014. Accessed May 12, 2017.
141. Rowland IR. Factors affecting metabolic activity of the intestinal microflora. *Drug Metab Rev.* 1988;19(3-4):243-261.
142. Fagherazzi G, Vilier A, Saes Sartorelli D, Lajous M, Balkau B, Clavel-Chapelon F. Consumption of artificially and sugar-sweetened beverages and incident type 2 diabetes in the Etude Epidemiologique aupres des femmes de la Mutuelle Generale de l'Education Nationale-European Prospective Investigation into Cancer and Nutrition cohort. *Am J Clin Nutr.* 2013;97(3):517-523.
143. Azad MB, Abou-Setta AM, Chauhan BF, et al. Nonnutritive sweeteners and cardiometabolic health: a systematic review and meta-analysis of randomized controlled trials and prospective cohort studies. *CMAJ.* 2017;189(28):E929-E939.
144. Suez J, Korem T, Zeevi D, et al. Artificial sweeteners induce glucose intolerance by altering the gut microbiota. *Nature.* 2014;514(7521):181-186.
145. Barber LK, Taylor SG, Burton JP, Bailey SF. A self-regulatory perspective of work-to-home undermining spillover/crossover: examining the roles of sleep and exercise. *J Appl Psychol.* 2017;102(5):753-763.
146. Hanlon EC, Tasali E, Leproult R, et al. Sleep restriction enhances the daily rhythm of circulating levels of endocannabinoid 2-arachidonoylglycerol. *Sleep.* 2016;39(3):653-664.

147. Nedeltcheva AV, Kilkus JM, Imperial J, Kasza K, Schoeller DA, Penev PD. Sleep curtailment is accompanied by increased intake of calories from snacks. *Am J Clin Nutr.* 2009;89(1):126-133.
148. Stickgold R, Walker MP. Memory consolidation and reconsolidation: what is the role of sleep? *Trends Neurosci.* 2005;28(8):-408-415.
149. Wagner U, Gais S, Haider H, Verleger R, Born J. Sleep inspires insight. *Nature.* 2004;427(6972):352-355.
150. Mander BA, Winer JR, Walker MP. Sleep and human aging. *Neuron.* 2017;94(1):19-36.
151. Walker MP, Robertson EM. Memory processing: ripples in the resting brain. *Curr Biol.* 2016;26(6):R239-R241.
152. Iliff JJ, Wang M, Liao Y, et al. A paravascular pathway facilitates CSF flow through the brain parenchyma and the clearance of interstitial solutes, including amyloid beta. *Sci Transl Med.* 2012;4(147):147ra111.
153. Iliff JJ, Lee H, Yu M, et al. Brain-wide pathway for waste clearance captured by contrast-enhanced MRI. *J Clin Invest.* 2013;123(3):1299-1309.
154. Xie L, Kang H, Xu Q, et al. Sleep drives metabolite clearance from the adult brain. *Science.* 2013;342(6156):373-377.
155. Spiegel K, Tasali E, Penev P, Van Cauter E. Brief communication: sleep curtailment in healthy young men is associated with decreased leptin levels, elevated ghrelin levels, and increased hunger and appetite. *Ann Intern Med.* 2004;141(11):846-850.
156. Schernhammer ES, Laden F, Speizer FE, et al. Rotating night shifts and risk of breast cancer in women participating in the Nurses' Health Study. *J Natl Cancer Inst.* 2001;93(20):1563-1568.
157. Tucker P, Marquie JC, Folkard S, Ansiau D, Esquirol Y. Shiftwork and metabolic dysfunction. *Chronobiol Int.* 2012;29(5):549-555.
158. Alegria-Torres JA, Baccarelli A, Bollati V. Epigenetics and lifestyle. *Epigenomics.* 2011;3(3):267-277.
159. Chaix A, Zarrinpar A, Miu P, Panda S. Time-restricted feeding is a preventative and therapeutic intervention against diverse nutritional challenges. *Cell Metab.* 2014;20(6):991-1005.

160. Zarrinpar A, Chaix A, Yooseph S, Panda S. Diet and feeding pattern affect the diurnal dynamics of the gut microbiome. *Cell Metab.* 2014;20(6):1006-1017.
161. American Psychological Association. Stress in America: paying with our health. https://www.apa.org/news/press/releases/stress/2014/stress-report.pdf. Published February 4, 2015. Accessed May 15, 2017.
162. Selye H. A syndrome produced by diverse nocuous agents. *Nature.* 1936;138:32.
163. Adam TC, Epel ES. Stress, eating and the reward system. *Physiol Behav.* 2007;91(4):449-458.
164. Adam TC, Tsao S, Page KA, Hu H, Hasson RE, Goran MI. Insulin sensitivity and brain reward activation in overweight Hispanic girls: a pilot study. *Pediatr Obes.* 2015;10(1):30-36.
165. Luo S, Romero A, Adam TC, Hu HH, Monterosso J, Page KA. Abdominal fat is associated with a greater brain reward response to high-calorie food cues in Hispanic women. *Obesity.* 2013;21(10):2029-2036.
166. Fava M, Judge R, Hoog SL, Nilsson ME, Koke SC. Fluoxetine versus sertraline and paroxetine in major depressive disorder: changes in weight with long-term treatment. *J Clin Psychiatry.* 2000;61(11):863-867.
167. Paykel ES, Mueller PS, De la Vergne PM. Amitriptyline, weight gain and carbohydrate craving: a side effect. *Br J Psychiatry.* 1973;123(576):501-507.
168. Aquila R, Emanuel M. Weight gain and antipsychotic medications. *J Clin Psychiatry.* 1999;60(5):336-337.
169. Casey DE. Side effect profiles of new antipsychotic agents. *J Clin Psychiatry.* 1996;57(suppl 11):40-45; discussion 46-52.
170. Smith SR, De Jonge L, Volaufova J, Li Y, Xie H, Bray GA. Effect of pioglitazone on body composition and energy expenditure: a randomized controlled trial. *Metabolism.* 2005;54(1):24-32.
171. Decanay EP, Van Metre TE. Suppression of asthma, weight gain, and linear growth in children receiving fluprednisolone. *J Allergy.* 1962;33:259-270.

172. Ratliff JC, Barber JA, Palmese LB, Reutenauer EL, Tek C. Association of prescription H1 antihistamine use with obesity: results from the National Health and Nutrition Examination Survey. *Obesity.* 2010;18(12):2398-2400.
173. Ponterio E, Gnessi L. Adenovirus 36 and obesity: an overview. *Viruses.* 2015;7(7):3719-3740.
174. Lin WY, Dubuisson O, Rubicz R, et al. Long-term changes in adiposity and glycemic control are associated with past adenovirus infection. *Diabetes Care.* 2013;36(3):701-707.
175. Lawless K. *Formerly Known as Food: How the Industrial Food System Is Changing Our Minds, Bodies, and Culture.* New York: St. Martin's Press; forthcoming.
176. An R. Fast-food and full-service restaurant consumption and daily energy and nutrient intakes in US adults. *Eur J Clin Nutr.* 2016;70(1):97-103.
177. Food and Drugs. Title 21 Code of Federal Regulations. §170.30-(c) and 170.3(f) (2016).
178. Environmental Working Group. EWG's Dirty Dozen guide to food additives. http://www.ewg.org/research/ewg-s-dirty-dozen-guide-food-additives#.WaFST1GQzMA. Published November 12, 2014. Accessed May 15, 2017.
179. McEvoy M. Organic 101: what the USDA Organic label means. In: US Department of Agriculture, Health and Safety. https://www.usda.gov/media/blog/2012/03/22/organic-101-what-usda-organic-label-means. Published March 22, 2012. Accessed July 10, 2017.
180. Baranski M, Srednicka-Tober D, Volakakis N, et al. Higher antioxidant and lower cadmium concentrations and lower incidence of pesticide residues in organically grown crops: a systematic literature review and meta-analyses. *Br J Nutr.* 2014;112(5): 794-811.
181. Smith-Spangler C, Brandeau ML, Hunter GE, et al. Are organic foods safer or healthier than conventional alternatives?: a systematic review. *Ann Intern Med.* 2012;157(5):348-366.
182. Srednicka-Tober D, Baranski M, Seal CJ, et al. Higher PUFA and n-3 PUFA, conjugated linoleic acid, alpha-tocopherol and

iron, but lower iodine and selenium concentrations in organic milk: a systematic literature review and meta- and redundancy analyses. *Br J Nutr.* 2016;115(6):1043-1060.

183. Sustainable Food Trust. http://www.sustainablefoodtrust.org.
184. Whoriskey P. The labels said "organic." But these massive imports of corn and soybeans weren't. *Washington Post.* https://www.washingtonpost.com/business/economy/the-labels-said-organic-but-these-massive-imports-of-corn-and-soybeans-werent/2017/05/12/6d165984-2b76-11e7-a616-d7c8a68c1a66_story.html?utm_term=.0ca05d41940f. Published May 12, 2017. Accessed September 25, 2017.
185. Organic Trade Association. OTA takes action on fraudulent imports. https://ota.com/news/issues/ota-takes-action-fraudulent-imports. Accessed September 25, 2017.
186. Thill S. Audit reveals weaknesses in USDA organic program oversight. http://civileats.com/2017/09/25/audit-reveals-weaknesses-in-usda-organic-program-oversight/. Published September 25, 2017. Accessed September 25, 2017.
187. FoodExpert-ID: the identity card. http://www.gene-express.de/content/C4_en.html.
188. Environmental Working Group. Dirty Dozen: EWG's 2017 shopper's guide to pesticides in produce. https://www.ewg.org/foodnews/dirty_dozen_list.php. Accessed May 16, 2017.
189. Environmental Working Group. Clean Fifteen: EWG's 2017 shopper's guide to pesticides in produce. https://www.ewg.org/foodnews/clean_fifteen_list.php. Accessed May 16, 2017.
190. Gillam C. *Whitewash: The Story of a Weed Killer, Cancer, and the Corruption of Science.* Washington, DC: Island Press; 2017.
191. Myers JP, Antoniou MN, Blumberg B, et al. Concerns over use of glyphosate-based herbicides and risks associated with exposures: a consensus statement. *Environ Health.* 2016;15:19.
192. Benbrook CM. Trends in glyphosate herbicide use in the United States and globally. *Environ Sci Eur.* 2016;28(1):3.
193. Environmental Working Group. EWG's Tap Water Database. https://www.ewg.org/tapwater/#.WY0QxyZK3fE. Accessed August 10, 2017.

194. International Bottled Water Association. Bottled water market. http://www.bottledwater.org/economics/bottled-water-market. Accessed May 13, 2017.
195. Rosinger A, Herrick K, Gahche J, Park S. Sugar-sweetened beverage consumption among U.S. youth, 2011–2014. *NCHS Data Brief No 271.* Hyattsville, MD: National Center for Health Statistics; 2017.
196. Rosinger A, Herrick K, Gahche J, Park S. Sugar-sweetened beverage consumption among U.S. adults, 2011–2014. *NCHS Data Brief No 270.* Hyattsville, MD: National Center for Health Statistics; 2017.
197. Pollan M. *In Defense of Food: An Eater's Manifesto.* New York: Penguin Press; 2008.
198. Sacks FM, Lichtenstein AH, Wu JHY, et al. Dietary fats and cardiovascular disease: a presidential advisory from the American Heart Association. *Circulation.* 2017;136(8).
199. Mahan B. Bayer and LibertyLink soybeans help protect hearts in America's heartland. https://www.cropscience.bayer.us/news/press-releases/2017/03022017-bayer-and-libertylink-soybeans-help-protect-hearts-in-americas-heartland. Crop Science, a division of Bayer. Published March 2, 2017. Accessed July 2, 2017.
200. Taubes G. Vegetable oils, (Francis) Bacon, Bing Crosby, and the American Heart Association. http://www.cardiobrief.org/2017/06/16/guest-post-vegetable-oils-francis-bacon-bing-crosby-and-the-american-heart-association. Published June 16, 2017. Accessed July 2, 2017.
201. Garaulet M, Gómez-Abellán P, Alburquerque-Béjar JJ, Lee YC, Ordovás JM, Scheer FA. Timing of food intake predicts weight loss effectiveness. *Int J Obes.* 2013;37(4):604-611.
202. Garaulet M, Vera B, Bonnet-Rubio G, Gómez-Abellán P, Lee YC, Ordovás JM. Lunch eating predicts weight-loss effectiveness in carriers of the common allele at PERILIPIN1: the ONTIME (Obesity, Nutrigenetics, Timing, Mediterranean) study. *Am J Clin Nutr.* 2016;104(4):1160-1166.
203. Lavers JL, Bond AL. Exceptional and rapid accumulation of anthropogenic debris on one of the world's most remote and

pristine islands. *Proc Natl Acad Sci U S A*. 2017;114(23):6052-6055.

204. Teeguarden J, Hanson-Drury S, Fisher JW, Doerge DR. Are typical human serum BPA concentrations measurable and sufficient to be estrogenic in the general population? *Food Chem Toxicol*. 2013;62:949-963.
205. Boucher JG, Boudreau A, Ahmed S, Atlas E. In vitro effects of bisphenol A beta-D-glucuronide (BPA-G) on adipogenesis in human and murine preadipocytes. *Environ Health Perspect*. 2015;123(12):1287-1293.
206. vom Saal FS, Cooke PS, Buchanan DL, et al. A physiologically based approach to the study of bisphenol A and other estrogenic chemicals on the size of reproductive organs, daily sperm production, and behavior. *Toxicol Ind Health*. 1998;14(1-2):239-260.
207. Howdeshell KL, Hotchkiss AK, Thayer KA, Vandenbergh JG, vom Saal FS. Exposure to bisphenol A advances puberty. *Nature*. 1999;401(6755):763-764.
208. Nagel SC, vom Saal FS, Thayer KA, Dhar MG, Boechler M, Welshons WV. Relative binding affinity-serum modified access (RBA-SMA) assay predicts the relative in vivo bioactivity of the xenoestrogens bisphenol A and octylphenol. *Environ Health Perspect*. 1997;105(1):70-76.
209. Nagel SC, vom Saal FS, Welshons WV. Developmental effects of estrogenic chemicals are predicted by an in vitro assay incorporating modification of cell uptake by serum. *J Steroid Biochem Mol Biol*. 1999;69(1-6):343-357.
210. Palanza PL, Howdeshell KL, Parmigiani S, vom Saal FS. Exposure to a low dose of bisphenol A during fetal life or in adulthood alters maternal behavior in mice. *Environ Health Perspect*. 2002;110(suppl 3):415-422.
211. Welshons WV, Nagel SC, Thayer KA, Judy BM, vom Saal FS. Low-dose bioactivity of xenoestrogens in animals: fetal exposure to low doses of methoxychlor and other xenoestrogens increases adult prostate size in mice. *Toxicol Ind Health*. 1999;15(1-2):12-25.

212. Hileman B. Conflict of interest: consulting firm assisting review of bisphenol A risks also works for compound's manufacturers. *Chem Eng News*. 2007;85(11):13.
213. Rochester JR, Bolden AL. Bisphenol S and F: a systematic review and comparison of the hormonal activity of bisphenol A substitutes. *Environ Health Perspect*. 2015;123(7):643-650.
214. Catanese MC, Vandenberg LN. Bisphenol S (BPS) alters maternal behavior and brain in mice exposed during pregnancy/lactation and their daughters. *Endocrinology*. 2017;158(3):516-530.
215. Palanza P. The "plastic" mother. *Endocrinology*. 2017;158(3): 461-463.
216. Wagner M, Schlusener MP, Ternes TA, Oehlmann J. Identification of putative steroid receptor antagonists in bottled water: combining bioassays and high-resolution mass spectrometry. *PLoS One*. 2013;8(8):e72472.
217. Oreskes N, Conway EM. *Merchants of Doubt: How a Handful of Scientists Obscured the Truth on Issues from Tobacco Smoke to Global Warming*. 1st US ed. New York: Bloomsbury Press; 2010.
218. Environmental Working Group. Body burden: the pollution in newborns. http://www.ewg.org/research/body-burden-pollution-newborns. Published July 14, 2005. Accessed May 13, 2017.
219. Environmental Working Group. Pollution in minority newborns: BPA and other cord blood pollutants. http://www.ewg.org/research/minority-cord-blood-report/bpa-and-other-cord-blood-pollutants. Published November 23, 2009. Accessed May 13, 2017.
220. Bern HA. Fragile fetus. In: Colborn T, Clement C, eds. *Chemically-Induced Alterations in Sexual and Functional Development: The Wildlife/Human Connection*. Princeton, NJ: Princeton Scientific Publishing; 1992:403.
221. Fisher M, Arbuckle TE, Liang CL, et al. Concentrations of persistent organic pollutants in maternal and cord blood from the maternal-infant research on environmental chemicals (MIREC) cohort study. *Environ Health*. 2016;15(1):59.
222. Shapiro GD, Dodds L, Arbuckle TE, et al. Exposure to organophosphorus and organochlorine pesticides, perfluoroalkyl substances,

and polychlorinated biphenyls in pregnancy and the association with impaired glucose tolerance and gestational diabetes mellitus: the MIREC Study. *Environ Res.* 2016;147:71-81.

223. Norrgran Engdahl J, Bignert A, Jones B, Athanassiadis I, Bergman A, Weiss JM. Cats' internal exposure to selected brominated flame retardants and organochlorines correlated to house dust and cat food. *Environ Sci Technol.* 2017;51(5):3012-3020.
224. Mitro SD, Dodson RE, Singla V, et al. Consumer product chemicals in indoor dust: a quantitative meta-analysis of U.S. studies. *Environ Sci Technol.* 2016;50(19):10661-10672.
225. Fang M, Webster TF, Stapleton HM. Effect-directed analysis of human peroxisome proliferator-activated nuclear receptors (PPARγ1) ligands in indoor dust. *Environ Sci Technol.* 2015;49(16):10065-10073.
226. Fang M, Webster TF, Stapleton HM. Activation of human peroxisome proliferator-activated nuclear receptors (PPARγ1) by semi-volatile compounds (SVOCs) and chemical mixtures in indoor dust. *Environ Sci Technol.* 2015;49(16):10057-10064.
227. Environmental Working Group. EWG's Skin Deep cosmetics database. http://www.ewg.org/skindeep. Accessed May 16, 2017.
228. Geer LA, Pycke BF, Waxenbaum J, Sherer DM, Abulafia O, Halden RU. Association of birth outcomes with fetal exposure to parabens, triclosan and triclocarban in an immigrant population in Brooklyn, New York. *J Hazard Mater.* 2017;323(pt A):177-183.
229. Environmental Working Group. EWG's guide to healthy cleaning. http://www.ewg.org/guides/cleaners. Accessed May 16, 2017.
230. Whelan J. *Stink!* https://www.stinkmovie.com. Released November 27, 2015.
231. Kannan K, Takahashi S, Fujiwara N, Mizukawa H, Tanabe S. Organotin compounds, including butyltins and octyltins, in house dust from Albany, New York, USA. *Arch Environ Contam Toxicol.* 2010;58(4):901-907.
232. Body Worlds. http://www.bodyworlds.com/en.html.
233. Leary SD, Smith GD, Rogers IS, Reilly JJ, Wells JC, Ness AR. Smoking during pregnancy and offspring fat and lean mass in childhood. *Obesity.* 2006;14(12):2284-2293.

234. US Department of Health & Human Services/BeTobacco Free.gov. Tobacco facts and figures. https://betobaccofree.hhs.gov/about-tobacco/facts-figures/index.html. Accessed May 14, 2017.
235. Airports Council International. Aircraft movements for the past 12 months, ending Dec 2015. http://www.aci.aero/Data-Centre/Monthly-Traffic-Data/Aircraft-Movements/12-months. Updated April 11, 2016. Accessed May 10, 2017.
236. Hudda N, Gould T, Hartin K, Larson TV, Fruin SA. Emissions from an international airport increase particle number concentrations 4-fold at 10 km downwind. *Environ Sci Technol.* 2014;48(12):6628-6635.
237. McConnell R, Gilliland FD, Goran M, Allayee H, Hricko A, Mittelman S. Does near-roadway air pollution contribute to childhood obesity? *Pediatr Obes.* 2016;11(1):1-3.
238. Jerrett M, McConnell R, Wolch J, et al. Traffic-related air pollution and obesity formation in children: a longitudinal, multilevel analysis. *Environ Health.* 2014;13:49.
239. McConnell R, Shen E, Gilliland FD, et al. A longitudinal cohort study of body mass index and childhood exposure to secondhand tobacco smoke and air pollution: the Southern California Children's Health Study. *Environ Health Perspect.* 2015;123(4):360-366.
240. Hansell AL, Blangiardo M, Fortunato L, et al. Aircraft noise and cardiovascular disease near Heathrow airport in London: small area study. *BMJ.* 2013;347:f5432.
241. Correia AW, Peters JL, Levy JI, Melly S, Dominici F. Residential exposure to aircraft noise and hospital admissions for cardiovascular diseases: multi-airport retrospective study. *BMJ.* 2013;347:f5561.
242. Munzel T, Gori T, Babisch W, Basner M. Cardiovascular effects of environmental noise exposure. *Eur Heart J.* 2014;35(13):829-836.
243. Theakston F. Burden of disease from environmental noise: quantification of healthy life years lost in Europe. World Health

Organization Regional Office for Europe, Joint Research Center/European Commission. http://www.euro.who.int/__data/assets/pdf_file/0008/136466/e94888.pdf. Published June 1, 2011. Accessed April 10, 2017.

244. Ohayon M, Wickwire EM, Hirshkowitz M, et al. National Sleep Foundation's sleep quality recommendations: first report. *Sleep Health*. 2017;3(1):6-19.
245. Tosini G, Ferguson I, Tsubota K. Effects of blue light on the circadian system and eye physiology. *Mol Vis*. 2016;22:61-72.
246. Chang AM, Aeschbach D, Duffy JF, Czeisler CA. Evening use of light-emitting eReaders negatively affects sleep, circadian timing, and next-morning alertness. *Proc Natl Acad Sci U S A*. 2015;112(4):1232-1237.
247. Milne S, Elkins MR. Exercise as an alternative treatment for chronic insomnia (PEDro synthesis). *Br J Sports Med*. 2017;51(5):479.
248. Sleep in America Poll. 2013 exercise and sleep. National Sleep Foundation. https://sleepfoundation.org/sleep-polls-data/sleep-in-america-poll/2013-exercise-and-sleep. Published February 20, 2013. Accessed April 10, 2017.
249. Morgan K. Daytime activity and risk factors for late-life insomnia. *J Sleep Res*. 2003;12(3):231-238.
250. Owen N, Healy GN, Matthews CE, Dunstan DW. Too much sitting: the population health science of sedentary behavior. *Exerc Sport Sci Rev*. 2010;38(3):105-113.
251. Katzmarzyk PT, Lee IM. Sedentary behaviour and life expectancy in the USA: a cause-deleted life table analysis. *BMJ Open*. 2012;2(4).
252. Koepp GA, Moore GK, Levine JA. Chair-based fidgeting and energy expenditure. *BMJ Open Sport Exerc Med*. 2016;2(1):e000152.
253. Srikanthan P, Karlamangla AS. Relative muscle mass is inversely associated with insulin resistance and prediabetes: findings from the Third National Health and Nutrition Examination Survey. *J Clin Endocrinol Metab*. 2011;96(9):2898-2903.

254. Souder W. *A Plague of Frogs: The Horrifying True Story.* 1st ed. New York: Hyperion; 2000.
255. Kristof N. Warnings from a flabby mouse. *New York Times.* http://www.nytimes.com/2013/01/20/opinion/sunday/kristof-warnings-from-a-flabby-mouse.html. Published January 19, 2013. Accessed September 23, 2017.

Image Credits

Illustration on page 13 courtesy of Eric Jay Decetis. Photo on page 18 courtesy of Shutterstock. Illustration on page 27 courtesy of James Janesick. Image on page 47 from the UNEP/WHO State of the Science of Endocrine Disruptors 2012 report. Illustration on page 71 courtesy of Geoff Olson. Photo on page 79 courtesy of Retha Newbold.

Index

About the Authors

Dr. Bruce Blumberg has been conducting pioneering research in the field of human biology for more than thirty years. He is professor in the Departments of Developmental and Cell Biology, Pharmaceutical Sciences, and Biomedical Engineering at the University of California–Irvine (UCI), which he joined in 1998. At UCI, his laboratory studies the biology of nuclear hormone receptors in development, physiology, and disease with a particular emphasis on how these are affected by hormonally active compounds in the diet and environment. He is credited with coining the term "obesogens" and proposing the obesogen hypothesis, which holds that exposure to environmental chemicals that disrupt the hormonal system may be an important factor predisposing individuals to weight gain and obesity. (Google the word "obesogens" and Dr. Blumberg is close by.) His laboratory demonstrated that the effects of maternal obesogen exposure may be passed on to future generations, as far as four generations out.

As one of the most respected biologists and scientific researchers in the world given his immense contributions to academic literature, Dr. Blumberg and his work have not gone unnoticed. Ever since "obesogens" landed in the mainstream media through articles in the *New York Times*, *US News & World Report*, *Atlantic*,

Washington Post, *Shape*, and even *The Dr. Oz Show*, Dr. Blumberg's phone has been ringing off the hook. He is known locally, nationally, and internationally as one of the top go-to experts in his specialty, especially with regard to understanding obesogens and how to avoid them. His knowledge is sought by not only journalists and health bloggers, but also other health professionals and obesity researchers. He has long intended to write this book, but waited patiently for the science to speak for itself. And now it has.

A graduate of Rutgers University ('76), Dr. Blumberg earned his PhD in biology from UCLA in 1987 and went on to make numerous discoveries in high-profile laboratories, including the prestigious Gene Expression Laboratory of Dr. Ron Evans at the Salk Institute for Biological Studies in La Jolla, California. To date, Dr. Blumberg has developed six U.S. patents and has published more than 137 peer-reviewed research papers that have been cited more than twenty-four thousand times. Dr. Blumberg was elected as a fellow of the American Association for the Advancement of Science (AAAS) in 2012.

When he is not immersed in his work, Dr. Blumberg enjoys being at home with his wife or hanging out at the stables watching his daughter ride horses. He has kept aquariums all his life with a current obsession with reef aquariums. He also loves digital photography and building computers. Dr. Blumberg is excited to share what he has learned from behind the scenes in his laboratory with the world.

Kristin Loberg specializes in taking experts' knowledge and turning it into reader-friendly works for general audiences. Kristin has a lengthy list of successful collaborations; multiple books on which she's worked have spent weeks on the *New*

York Times, *Wall Street Journal*, Amazon, and *USA Today* best-seller lists. She earned her degree from Cornell University and lives in Los Angeles where she was born. She is a member of the Author's Guild and PEN, and she occasionally teaches an intensive proposal-writing workshop at UCLA.